Michael Rieger

First-principles based models for lateral interactions of adsorbates

Michael Rieger

First-principles based models for lateral interactions of adsorbates

A route towards more realistic models for multiscale simulations of heterogeneous catalysis

Südwestdeutscher Verlag für Hochschulschriften

Impressum/Imprint (nur für Deutschland/only for Germany)
Bibliografische Information der Deutschen Nationalbibliothek: Die Deutsche Nationalbibliothek verzeichnet diese Publikation in der Deutschen Nationalbibliografie; detaillierte bibliografische Daten sind im Internet über http://dnb.d-nb.de abrufbar.
Alle in diesem Buch genannten Marken und Produktnamen unterliegen warenzeichen-, marken- oder patentrechtlichem Schutz bzw. sind Warenzeichen oder eingetragene Warenzeichen der jeweiligen Inhaber. Die Wiedergabe von Marken, Produktnamen, Gebrauchsnamen, Handelsnamen, Warenbezeichnungen u.s.w. in diesem Werk berechtigt auch ohne besondere Kennzeichnung nicht zu der Annahme, dass solche Namen im Sinne der Warenzeichen- und Markenschutzgesetzgebung als frei zu betrachten wären und daher von jedermann benutzt werden dürften.

Verlag: Südwestdeutscher Verlag für Hochschulschriften GmbH & Co. KG
Heinrich-Böcking-Str. 6-8, 66121 Saarbrücken, Deutschland
Telefon +49 681 37 20 271-1, Telefax +49 681 37 20 271-0
Email: info@svh-verlag.de

Approved by: Berlin, FU, Dissertation 2010

Herstellung in Deutschland:
Schaltungsdienst Lange o.H.G., Berlin
Books on Demand GmbH, Norderstedt
Reha GmbH, Saarbrücken
Amazon Distribution GmbH, Leipzig
ISBN: 978-3-8381-2903-7

Imprint (only for USA, GB)
Bibliographic information published by the Deutsche Nationalbibliothek: The Deutsche Nationalbibliothek lists this publication in the Deutsche Nationalbibliografie; detailed bibliographic data are available in the Internet at http://dnb.d-nb.de.
Any brand names and product names mentioned in this book are subject to trademark, brand or patent protection and are trademarks or registered trademarks of their respective holders. The use of brand names, product names, common names, trade names, product descriptions etc. even without a particular marking in this works is in no way to be construed to mean that such names may be regarded as unrestricted in respect of trademark and brand protection legislation and could thus be used by anyone.

Publisher: Südwestdeutscher Verlag für Hochschulschriften GmbH & Co. KG
Heinrich-Böcking-Str. 6-8, 66121 Saarbrücken, Germany
Phone +49 681 37 20 271-1, Fax +49 681 37 20 271-0
Email: info@svh-verlag.de

Printed in the U.S.A.
Printed in the U.K. by (see last page)
ISBN: 978-3-8381-2903-7

Copyright © 2011 by the author and Südwestdeutscher Verlag für Hochschulschriften GmbH & Co. KG and licensors
All rights reserved. Saarbrücken 2011

Abstract

The adsorption of a particle or molecule from the gas-phase is what is typically referred to as the first step in heterogeneously catalyzed processes. The adsorbed particles are activated for the subsequent chemical reaction by the chemical bond being formed between the particle and the specific adsorption site on the surface. How reactive an adsorbed particle will be is strongly influenced by the strength of this bond to the surface. A strong bonding to the surface can lead to a low reactivity. In case of a weak bond, it is possible that the particle desorbs again before a reaction could take place. This bond strength is not only influenced by the formation energy of the chemical bond to the surface but also by the interactions between neighboring particles.

These interactions can have either a stabilizing or destabilizing effect. Their importance increases the more densely packed the adsorbates are on the surface. The higher the coverage, the more important the impact of these interactions. Ordering behavior or even the promotion of certain cluster or island structures are two examples for properties which are steered by these interactions on a microscopic level. Quantifying these effects on a microscopic scale is not always an easy task.

It is, therefore, of great importance for model descriptions which are used in computer simulations for these kind of systems to properly describe the interactions of the adsorbates in the plane parallel to the surface — the *lateral* interactions. A major motivation for the development of such model descriptions is presented by the fact that even nowadays supercomputers are not powerful enough to simulate, on the level of electronic structure theory, system sizes large enough for a realistic description.

An ideal model would be presented by one whose parameters are derived from a set of calculations on the level of electronic structure theory. In addition, the derived parameters of this model should be chosen such that even in the mesoscopic model all parameters would retain their microscopic meaning.

This is the major topic of the thesis presented here. This work presents an important contribution because it takes a systematic look at this important interplay between different kinds of interaction and derives appropriate model descriptions for the simulation of large scale systems. Starting from a simple qualitative model for the description of the interplay between lateral interactions and chemisorption, two limiting cases will be discussed.

In the first case a strong bond to the specific adsorption site is established and neighboring particles interact only weakly. This scenario will be investigated throughout this thesis by means of a cluster expansion (CE) approach. Based on calculations on the level of density-functional theory a parameterization of interaction parameters will be presented. In this model description lateral interaction parameters are identified with a certain geometrical configuration of neighboring particles adsorbed in surface adsorption sites. Within the parameterization process a certain energy contribution to the total binding energy of a particle will be assigned to this specific adsorption configuration. By means of this parameterization procedure it is ensured that all interaction parameters in fact resemble a microscopic arrangement of particles. This approach will be applied to a classical example system of surface science. It will be used to describe the adsorption of carbon monoxide (CO) on a palladium (100) surface. As a result of this application a closed set of interaction parameters will be derived which is in very good agreement with previously published values of certain interaction parameters. These previously published values, in contrast, had been derived by a fitting process to experimental data. Limitations of this description will be discussed for the specific example system and their origins will be explained.

The second case presents an example of strongly interacting adsorbates. The binding to the surface is either rather weak or at most of the same order of magnitude as the particle interactions. In such a scenario usually a non site specific adsorption is observed. A typical example is presented by a system where a metal is deposited on a surface of another metal. In the work presented here, the example system is presented by small copper islands on a silver surface. Motivated by a set of experimental observations for which an explanation was desired, this system was investigated by a theoretical approach. Based on simulations on the level of electronic structure theory a semi-empirical potential description of the example system was validated. This potential was then used for large scale simulations accessing the experimental system sizes. The insight gained by our simulations allowed us to develop a model which is able to explain the driving force behind the experimentally observed reconstructions. We showed that the driving force is originated in the lateral interactions of island atoms and the strain induced by the large lattice mismatch of silver and copper. Our results were able to initiate new experiments. In addition, based on our simulations, the required experimental parameters could be suggested.

Zusammenfassung

Die Adsorption eines Teilchens oder Moleküls aus der Gasphase ist typischerweise der erste Schritt in heterogen katalysierten Prozessen. Hierbei werden die Teilchen durch die Ausbildung einer Bindung zur Oberfläche für die nachfolgende chemische Reaktion aktiviert. Die Reaktivität des adsorbierten Teilchens wird dabei maßgeblich durch die Bindungsenergie beeinflußt. Ist das Adsorbat zu stark an den Bindungsplatz der Oberfäche gebunden wir die Reaktion gehemmt. Liegt nur eine sehr schwache Bindung zur Oberfläche vor, so kann es sein, daß das Teilche desorbiert bevor überhaupt eine Reaktion stattfinden kann. Einflußgrößen auf die Bindungsenergie sind nicht nur die bei der Chemisorption des Teilchens freiwerdende Bindungsenergie für den spezifischen Bindungsplatz auf der Oberfläche sondern auch die Wechselwirkungen mit benachbarten Teilchen.

Diese Wechselwirkung kann stabilisierenden oder destabilisierenden Einfluß haben. Diese Wechselwirkungen werden wichtiger je dichter die Adsorbate angeordnet sind, je höher die Bedeckung einer Oberfläche umso stärker ist also der Einfluß dieser Wechselwirkungen. Sie bestimmen auf einem mikroskopischen Level zum Beispiel das Anordnungsverhalten zu geordneten Oberflächenstrukturen oder die Ausbildung bestimmter Cluster oder Inselstrukturen. Diese Wechselwirkungen zu quantifizieren ist nicht immer eine einfache Aufgabe

In Modellbeschreibungen solcher Systeme, die in Computersimulationen eingesetzt werden, ist es daher von entscheidender Bedeutung die Wechselwirkungen zwischen den Adsorbaten in der Ebene parallel zur Oberfläche — die *lateralen* Wechselwirkungen — richtig zu beschreiben und wiederzugeben. Die Notwendigkeit für die Aufstellung solcher Modelle ist in diesem Zusammenhang dadurch gegeben, daß für die Berechnung von Systemen in realistische Größenordnungen auf dem Niveau von Elektronenstrukturrechnungen auch moderne Supercomputer nicht ausreichen.

Im Idealfall sind solche Modellbeschreibungen von Rechnungen auf Elektronenstrukturrechnungen parametrisiert und die entscheidenden Parameter des mesoskopischen Modells sind so gewählt, daß sie in Verbindung zur mikroskopischen Realität stehen.

Die vorliegende Dissertation befaßt sich mit diesem Thema und liefert einen wichtigen Beitrag durch eine systematische Betrachtung dieses Zusammenspiels von Wechselwirkungen und einer Ableitung entsprechender Modellbeschreibungen für Computersimulationen großer Systeme. Ausgehend von einem einfachen qualitativen Modell zur Beschreibung des Zusammenspiels von lateralen Wechselwirkungen und Chemisorption an spezifischen Adsorptionsplätzen einer Oberfläche werden zwei Grenzfälle diskutiert.

Im ersten Fall überwiegt die Stärke der Bindung zum Adsorptionsplatz und die benachbarten Teilchen wechselwirken im Vergleich dazu schwach miteinander. Dieses Szenario wird im Rahmen dieser Arbeit mit einem Cluster Expansion (CE) Ansatz untersucht, bei aus Rechnungen auf dem Niveau der Dichtefunktionaltheorie Wechselwirkungsparameter parametrisiert werden. In diesem Modell stellt ein Wechselwirkungsparameter eine bestimmt geometrische Anordnung benachbarter Teilchen auf der Oberfläche dar, und die Parametrisierung weist dieser Anordnung einen bestimmten Energiebeitrag zur gesamten Bindungsenergie des adsorbierten Teilchens zu. In dieser Vorgehensweise behalten alle Modellparameter ihre mikroskopische Bedeutung. Dieser Ansatz wird an einem klassischen Beispielsystem der Oberflächenphysik, der Adsorption von Kohlenstoffmonoxid (CO) auf einer Palladium (100) Oberfläche eingesetzt. Im Rahmen dieses Einsatzes wird ein geschlossenes Set von Wechselwirkungsparametern abgeleitet, das in sehr guter übereinstimmung mit bereits publizierten Parametern ist, die in einem Fit Prozess von experimentellen Strukturen her abgeleitet worden sind. Grenzen der Modellbeschreibung für das spezifische Beispielsystem werden aufgezeigt und ihre Ursachen diskutiert.

Für den zweiten Fall wird die Struktur der Adsorbate durch ihre sehr starken lateralen Wechselwirkungen bestimmt. Die Bindung zur Oberfläche ist schwach oder zumindest in der gleichen Größenordnung wie die Adsorbatwechselwirkungen. In diesem Szenario wird im allgemeinen keine so ortsfeste Bindung zur Oberfläche beobachtet. Ein typisches Beispiel für einen solchen Fall findet man bei Systemen, bei denen ein Metall auf der Oberfläche eines anderen Metalls abgeschieden wird. Die vorliegende Arbeit untersucht einen solchen Fall motiviert durch eine Fragestellung aus Experimenten für das System von Kupferinseln auf einer Silberoberfläche. Ausgehend von Elektronenstrukturrechnungen wird für große Systeme ein semi-empirisches Potential validiert und anschließend dafür benutzt Simulationen für die experimentellen Systemgrößen zu machen. Die im Rahmen dieser Untersuchung erzielten Ergebnisse erlaubten zum einen die experimentellen Befunde zu erklären und ein Modell zu entwickeln, das die Struktur der Inseln auf Grundlage der lateralen Wechselwirkungen der Inselatome untereinander erklären kann. Als weitere Triebkraft wurde der große Unterschied in den Gitterkonstanten von Kupfer und Silber identifiziert. Im weiteren legten die erzielten Ergebnisse dieser Arbeit dann die Grundlage für weitergehende experimentelle Untersuchungen für die neue experimentelle Bedingungen aus unseren Simulationen bestimmt werden konnten.

Contents

I	Introduction	1
1	**Interacting Adsorbates**	3
	1.1 Origins of Lateral Interactions	5
2	**Theoretical Perspective**	7
	2.1 The Original Frenkel-Kontorova Model	7
	2.2 Quantities Provided by Simulations	11

II	Methods	15
3	**Electronic Structure Theory**	17
	3.1 The Schrödinger Equation of a Many-Body System	18
	3.2 Density-Functional Theory	21
	3.3 Application to Periodic Systems, Plane Waves and Pseudopotentials	26
	3.4 STM Simulations – The Tersoff-Hamann Approximation	29
4	**Bridging Scales**	31
5	**Cluster Expansion**	35
	5.1 Total Energy in the Cluster Expansion Formalism	35
	5.2 Cross Validation	37
	5.3 Ground-State Validation	38
	5.4 Sources of Uncertainty – Accuracy Considerations	41
6	**Embedded-Atom-Method**	45
	6.1 Total Energy in the Embedded-Atom-Method	45

	6.2	The Parametrization of the EAM Potential for pure Silver	48
	6.3	The Cross-Potential for the Ag–Cu Interaction	50
	6.4	On Benefits and the Validation of the EAM Potential	51

7 System Optimization — 53

7.1	Structure Optimization by Energy Minimization	54
7.2	Structure Optimization by Global Optimization	55
7.3	The Stability of Optimized Structures	58

III Application and Results — 67

8 CO Adsorption on Pd(100) — 69

8.1	The Pd(100) Surface	69
8.2	CO Adsorption - Experiments and Theory	70
8.3	The Lateral Interactions of Adsorbed CO on Pd	74
8.4	Conclusions and Limitations	90

9 Small Cu Islands on Ag(100) — 93

9.1	Lateral Interaction in Metal Heteroepitaxy	94
9.2	The Copper-Silver Alloy System	94
9.3	Theoretical Investigation	98
9.4	Conclusions and Outlook	119

IV Conclusions — 121

10 Final Remarks and Future Steps — 123

| 10.1 Future Steps | 124 |

Appendix — 124

A Convergence Tests — 125

| A.1 Convergence Tests for the DFT Calculations | 125 |
| A.2 Final Computational Setup | 133 |

B Adsorption Structure Database for the Cluster Expansion — 135

| B.1 Reference Energies | 136 |
| B.2 DFT Structure Database | 137 |

C Overview of Calculated Cu Island Structures — 165

| C.1 Structures Calculated on a DFT Level | 165 |

C.2 DFT Geometries 166
C.3 Comparison of Energy Differences for Different Island Sizes 168

Bibliography 169

Part I

Introduction

Chapter 1
Interacting Adsorbates

ADSORPTION is the first and most elementary step for all reaction processes in heterogenous catalysis. It is the key step for what is normally seen as the activation of the educts. The species, unreactive in the gas-phase, will change into a more reactive state upon adsorption on a surface, thereby allowing reactions to happen and products to be formed at lower temperatures and lower pressures than required for the according gas-phase reaction. Another mechanism how the surface facilitates the reaction is by its property to bring the reacting species close to each other in a certain — and for this reaction necessary — geometric configuration which could be very unlikely to be formed during collision events in the gas-phase.

A closely connected quantity to the process of adsorption which then influences the catalytic activity of the adsorbed species is what will be called *binding energy* of the adsorbed species in the following. In the definition used in this thesis, this term includes contributions from the direct bonding to the surface and energetic contributions caused by the fact that neighboring particles might interact via various mechanisms with each other.

Will there be some interaction effects at all? What physical kind of interaction will happen for different species? On what energy scale will the interaction happen in comparison to the energy necessary to activate the desired reaction? Will there be ordering of adsorbates or the generation of domains or islands of adsorbates due to the adsorbate interaction? What influence do those interaction have upon the properties of the surface and its reactivity? Summarizing, one can state that such mutual interactions between adsorbed particles play an important role for numerous surface phenomena, e.g. melting, roughening, crystal and thin film growth, catalysis and corrosion.[1]

One goal of the work carried out in this thesis was to study the influence of such interactions between two adsorbates of fundamentally different kind on metal surfaces. By extracting meaningful parameters from calculations the physical origin and the impact of these interactions on certain properties of the many particle systems on the surfaces was studied by means of computer simulations. Within these simulations, several length scales were bridged employing different approaches for an appropriate model description of the particle interactions. In one example, a model description was completely parameterized from first-principles using a self-developed computational framework for the cluster expansion approach. The second example system was studied using an established semi-empirical potential description, which was checked for its applicability to the specific problem by a validation against calculations on the level of electronic structure theory. The two model descriptions were chosen to be most appropriate for the specific system under investigation. After the application to the example systems the two different model descriptions were critically discussed. This discussion tried to cover a methodological perspective and a critical analysis of the limitations of each model description.

Summarizing, it can be stated that this work on one hand had a methodological motivation, which is presented by the focus on a stringent and proper connection of the different levels of theory used in the parameterization to derive parameters of a coarse grained model from an underlying finer level. To learn about the effects of interactions on a mesoscopic scale (nanometers, picoseconds), we must first aim to have a sufficiently accurate description on the smaller scale and then develop an (ideally) seamless link from the small scale to the larger scale without introducing more errors by taking unchecked approximations. Our tool for the smallest — the electronic — scale was the framework of density-functional theory (DFT). In the description of the converged electronic structure of the complete simulation system, all interaction effects of the two systems studied were intrinsically included within the approximations underlying the DFT method.

However, the maximum system size that we were aiming for was far beyond the capabilities of DFT calculations. The maximum system size that can be treated by this most accurate method is naturally limited by the computational effort that one is able to invest into these calculations. It makes no sense to waste years of CPU time for the very accurate calculation of one single configuration, when the configurational space that needs to be covered to be able to predict ensemble averages, would have a dimension of the order of millions of configurations. As a consequence, if one is interested in information about more than a very limited number of configurations, computationally less demanding methods must be used.

As such methods, which are derived and properly linked to the electronic structure regime, are just emerging in the field of materials simulations, there are not so many standardized approaches yet. Therefore, a large part of this thesis will be devoted to the method description and development which is necessary to establish this stringent linkage between the electronic level of calculations and the development and parametrization of mesoscopic model descriptions (here: Embedded-Atom-Method and Cluster-Expansion). One focus will be laid on the validation of these coarse grained model descriptions against high level theoretical data. In addition, the propagation of errors from the fine to the coarse level will be a topic in the discussion of the limitations of these model descriptions.

In addition, this work had clearly also a materials science motivation. Obviously by means of our simulations, it was possible to learn a lot about the governing effects of adsorbate ordering or reconstruction on different scales. With this we were able to understand and explain experimental observations. The interplay of microscopic interactions of a large number of adsorbed particles can be crucial, for example, in catalytic processes or in growth processes. For a theoretical description one needs to describe the two contributions mentioned in the beginning, namely, the interaction between adsorbed particles and the direct interaction to the substrate. For these two contributions limiting regimes can be defined which will be explained from a theoretical perspective in the next chapter.

1.1 Origins of Lateral Interactions

When adsorption on a surface takes place and more and more adsorbates are accumulated on the surface, the binding energy per adsorbate as defined in section 2.2.3 equation Eq. 2.8 changes as a function of coverage. The adsorbates are forced upon increasing coverage to pack more and more densely, lowering the average distance amongst them. The physical origins of these interactions can be separated either based on the way they are transmitted (through-space or through-substrate)[2], or just based on their seperation distance[3]. Both classifications are of course connected.

At short distances, direct through-space adsorbate–adsorbate interactions dominate. They are caused by direct electronic interactions of electronic orbitals of the adsorbed atoms or molecules. This overlap of atomic or molecular wavefunctions on neighboring adspecies falls off rapidly and is therefore very short ranged [2–4]. This effect corresponds to the $1/r^{12}$ term in a Lennard-Jones[5,6] potential description[4], with r being the adspecies separation distance.

At larger distances interactions are predominantly indirect. Through-space interactions are mediated at such distances electrostatically. They are caused by the dipole–dipole interaction of the adsorbate species' dipole moments. This interaction decays with a term $1/r^3$. Without an electric field, this interaction is a nonoscillatory one, however, in the case of spectroscopy also oscillatory effects can be observed[7].

In addition on metallic surfaces, through-substrate interactions take also effect at long distances between adsorbates. The deformation of the substrate lattice leads to a nonoscillatory elastic interaction between adsorbates which also decays with a term $1/r^3$. This interaction involves the surface lattice and the induced strain effects on it and is therfore clearly through-substrate. Interaction between adsorbate electrons with the surface state electrons of the substrate, finally, can lead to an osciallatory long range interaction between adsorbates that is mediated by the substrate band structure. The adsorbed atom/molecule imposes its potential onto the host electrons, which is then screened by density oscillations (Friedel oscillations)[8].

Chapter 2
Theoretical Perspective

THIS chapter introduces the two regimes of interplay between interacting adsorbates on surfaces that will be investigated within this study. This will be done by describing a very basic but quite elucidating theoretical model. For this, one can start with a one-dimensional chain of pairwise interacting particles sitting in a periodic one-dimensional corrugated potential. This potential is merely a construct to describe the surface, however, it does not necessarily have to describe only a surface. In a situation like this, the equilibrium state of the system will be determined by the interplay of two competing interactions, namely the particle-particle interaction and the particle-surface interaction.

In addition to this educational example, some basic quantities used throughout the theoretical investigation in this thesis will be introduced and described. A quite general definition of lateral interactions will be presented and some of the physical origins of these interactions will be mentioned and described. Furthermore, a short paragraph will explain some point to be cautious about when investigating lateral interactions with computer codes that employ periodic boundary conditions.

2.1 The Original Frenkel-Kontorova Model

The basic physics of this model was originally published already in 1938[9,10] within the context of the theory of dislocations on metals. Yet, despite its simplicity the Frenkel-Kontoroa (FK) model provides a quite rich variety of applications[11]. Since then it has e.g. been successfully used to gain insight into the underlying physics of surface reconstruction[12] and ordering of a sub-monolayer film of adsorbate on a crystal surface[13]. An extension from the most simple $1D$ case to a $2D$ description provided explanations for heteroepitaxial growth[14,15] and even dynamic processes on surfaces like island diffusion[16] were successfully investigated by means of this model description.

The central theme of this thesis is nicely illustrated by following the Frenkel-Kontorova model for some limiting scenarios. The effects of different ratios of the determining parameters within this model will be discussed. Although being able to cover and illustrate qualitatively the basic physics which underlie a lot of surface phenomena, for a description of more sophisticated model systems a – closer to real scenarios and aiming for quantitative descriptions – higher level models are needed. For this reason, the basic FK model will be only used as an introductionary showcase, but then the main focus of this thesis will be laid on the development of more quantitative models, which include also many-body interactions in different ways. In addition, an extended FK model will be used later as a starting point for the investigation of an island reconstruction mechanism with the aim to gain useful insights into the governing kinds of interactions.

2.1.1 Limiting Scenarios of the One-dimensional Frenkel-Kontorova Model

In Figure 2.1 the most simple version of the FK model for an interacting one-dimensional chain of adsorbates placed on a corrugated periodic surface potential is presented. Both systems, surface (index s) and chain (index c), present two different typical separation lengths. The sinusoidal surface potential exhibits the period a_s, whereas the chain of atoms are separated by harmonic springs of length a_c. The springs have an elastic constant, k, which acts as a counter force if the particle postions relative to each other deviate from the equilibrium distance.

When the chain of particles approaches the surface, it will experience an attraction due to the potential wells of the surface potential. Obviously, the maximum amount of binding energy between a particle and the surface potential could be gained if each particle occupied one minimum of the surface potential. If every minimum is occupied by one particle a monolayer coverage situation is established. However, placing every atom in a surface minimum would stretch the springs between the atoms, and therefore this configuration would not correspond to a minimum energy configuration with respect to the chain of atoms.

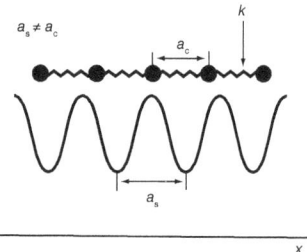

Figure 2.1: Schematic view of the one-dimensional Frenkel-Kontorova Model for the description of interacting adsorbates on a periodic surface potential.

On the other hand, if all atoms were in ideal equilibrium positions with respect to the harmonic springs, their position would differ significantly from those of the surface potential minima. The real equilibrium configuration will therefore lie somewhere in between these two situations, minimizing the total energy of the system under the constraints of the two competing potentials. For illustrative purposes let us assume a periodic surface potential of

2.1. The Original Frenkel-Kontorova Model

the form

$$V_{\text{misfit}} = \frac{A}{2}\left[\cos(2\pi x) - 1\right] \quad (2.1)$$

and an interaction between two chainatoms at x_i and x_{i+1} of

$$V_{\text{strain}} = k(x_i - x_{i+1} - a_c)^2/2 \quad (2.2)$$

The total energy of this system is the sum over the chain atoms:

$$E^{\text{total}} = \sum_{i=1,N} \frac{A}{2}\left[\cos(2\pi x) - 1\right] + \sum_{i=1,N-1} \frac{k}{2}(x_i - x_{i+1} - \varepsilon)^2 \quad , \quad (2.3)$$

with a_c and k as defined above, A is representing the depth of the surface potential well and N the total number of atoms. The equilibrium configuration of this simple system is mainly determined by a few factors:

i) The mismatch between a_c and a_s.
ii) The strength of the spring constant, k.
iii) The depth of the potential well A of the surface potential.

The spring constant k can easily be connected to the bond strength between the chain atoms, which in turn energetically can be related to the binding energy between chain atoms. Likewise, the depth of the potential well can be connected to the bond energy, E_b, of the adsorbate-surface bond.

Reducing the argumentation to the energetics, leaves us with two limiting situations:

1. The interaction between the chain atoms is much stronger than the bonding to the surface. That is for the characteristic lengths of the system:

$$k/a_c \geq E_b \cdot a_s \quad . \quad (2.4)$$

In this case, a so-called *floating adlayer* situation results, leading to a most likely incommensurate adlayer configuration of a layer of adsorbates loosely bound to the surface. Example cases for this situation are rare-gas adlayers on metal surfaces[17].

2. A scenario in which the adsorbates interact weakly with each other and each particle has a strong bond to the surface. Molecules adsorbed by chemisorption to the surface represent a showcase for this case. In such systems, it holds — using again the characteristic lengths introduced before:

$$k/a_c \leq E_b \cdot a_s \quad . \quad (2.5)$$

The adsorbates will occupy distinct adsorption sites in a so-called «site-specific» adsorption scenario, but the adsorbate–adsorbate interaction will determine the ordering behavior of the adsorbate lattice-gas in these sites for different coverages.

Unfortunately, real systems only very seldom fall into these two limiting categories and will very often exhibit a behavior that is governed to equal parts by the adsorbate–adsorbate interactions and the adsorbate–surface interactions. But nevertheless, this simple view outlined briefly and nicely the principle focus of this thesis.

2.1.2 Introducing the Example Systems Used

Heteroepitaxial crystal growth for a metal on metal system and ordering phenomena upon increasing coverage for a molecule on a metal surface are the two examples used in this work to discuss and present the different kinds of interactions that can present the underlying physics for certain experimental observations. These two scenarios are connected to the two limiting scenarios of the Frenkel-Kontorova model presented above by the ratio exhibited by the particle interaction and the particle–surface interaction, respectively.

For the limiting case presented under 1. above, the adsorption of carbon monoxide (CO) on the (100) surface of palladium, Pd, will be taken as a representative example. Chapter 8 will present the results that were gained using a cluster expansion (CE) framework to extract the relevant interaction parameters. A large time of the study presented in this thesis was invested into the development of a new and flexible computer code implementing the cluster expansion framework, which can be easily adapted to a large number of different surface–adsorbate systems by simple input files. As a benefit of this flexible structure, it was possible to validate the newly written code initially against the data and results of an earlier work performed in the group[18] and demonstrate its reliability in this way.

The second limiting case of the model will be studied in chapter 9. In this part the island formation of small copper, Cu, islands on a (100) surface of silver, Ag, will be examined motivated by unusual experimental findings in a recent STM study[19]. The approach taken will start from an extension of the simple Frenkel-Kontorova model. This extended model will be used in global optimization studies which goal will be to identify the most stable island structures. Based on the insights gained from the first-principles based parameterization of this extended Frenkel-Konorova model the study will proceed using a semi-empirical many-body potential, namely the embedded-atom-method (EAM), in a subsequent step. With the new potential description again the most stable island configuration for the reported experimental conditions will be investigated. Finally, based on the analysis of the simulations, a description for the observed size-dependent reconstruction behavior will be suggested.

Finally, these two different approaches for the description of interactions between adsorbates will be discussed and compared with respect to their underlying physical assumptions. Their limitations and strength for large scale simulations in a multiscale modeling environment will be briefly discussed in the conclusions.

2.2 Quantities Provided by Simulations

In theory and simulations, one deals with quantitative descriptors for certain phenomena. Here, shortly some fundamental quantities will be defined that are used throughout the thesis.

2.2.1 Total Energy

All methods used in this thesis are yielding first of all as fundamental quantity the *total energy*, E^{total} of a configuration of atoms. This total energy is calculated either by some potential description of the system, which relates usually the absolute particles position and the positions with respect to each other to an energy contribution, or by means of electronic structure theory, employing the principles of quantum mechanics.

Further quantities can be derived by taking into account, for example, first derivatives of the total energy with respect to particle position, yielding the acting forces on the respective particle. For electronic structure theory calculations a rich variety of properties can be derived by different methods using information about the electron density distribution in the system in addition to the energetics.

2.2.2 Surface Energy

Cutting a bulk crystal and generating a surface naturally changes the bond situation for the atoms that are now exposed in a certain direction to another phase than the bulk phase. Usually, this formation of a surface is accompanied by relaxation phenomena of the surface layers, that are driven by the forces resulting from this anisotropic bond environment. The equilibrium shape of the surface will have minimal forces and energetically, minimal surface energy per unit area, γ [1]. Often this quantity is just called *surface energy* and it is defined as surface excess energy per area of the particular crystal facet. Therefore, it describes the cost connected with the creation of a certain surface area.

The most stable surface will consequently be the one which minimizes γ and by means of that minimization endeavor taking influence on the resulting equilibrium shape of a crystal. This is the underlying principle of the famous *Wulff construction* for the prediction of equilibrium crystal shapes [20].

Most experimental data for the surface energy come from surface tension measurements in liquid phase, extrapolated to zero temperature, which includes a rather large uncertainty in the resulting values [21].

Within the supercell approach employed in this work, see section 2.2.4 below, the surface energy γ for $T = 0$ K of a clean surface can be calculated as

$$\gamma = \frac{1}{2A}\left(E^{\text{total}}_{\text{slab}} - N_{\text{surface}} \cdot E^{\text{total}}_{\text{bulk}}\right) \quad . \tag{2.6}$$

In equation (2.6), A denotes the surface unit area and the factor $\frac{1}{2}$ is caused by the fact that in the supercell approach the slab has two surfaces. The energetics of the complete slab and an atom in the bulk are given by $E_{\text{slab}}^{\text{total}}$ and $E_{\text{bulk}}^{\text{total}}$, respectively, while N_{surface} is the number of atoms in the slab.

For the description of a surface, γ is the quantity for which the convergence with respect to the calculation setup was tested in this thesis, see appendix A.

2.2.3 Binding Energy of Adsorbates - What do we mean when we talk about lateral interactions?

When a gas phase molecule or atom interacts with a surface, adsorption can take place and a bond is formed. This process releases a formation energy, that is usually referred to as *bonding energy*, E_{bond}. If now more than only one adsorbate adsorbs on a surface, the total formation energy is usually *not* a linear function of the coverage (where coverage θ is defined as $\theta = N_{\text{ads sites}}^{\text{occupied}} / N_{\text{ads sites}}^{\text{total}}$, the ratio between occupied and available adsorption sites). This is because the adsorbates, «ads», interact with each other in the surface plane via *lateral interactions*. These interactions can formally be seen as a certain energy contribution to the total formation energy that is additionally freed or taken by the interactions upon the adsorption process. Therefore, lateral interactions can be classified as repulsive or attractive forces between adsorbates which act in a plane parallel to the surface plane[4].

Within this thesis, the total formation energy for configurations with interacting adsorbates will be called *binding energy*, $E_{\text{bind}}^{\text{total}}$. In a first ansatz, one can write

$$E_{\text{bind}}^{\text{total}} = E_{\text{interaction}} + \sum_{i=1}^{N} E_{\text{bond}} \qquad (2.7)$$

This energy is a function of coverage and adsorbate distribution. In a potential description which is yielding the total energy of a given configuration (including implicitly also all interactions in addition to the bonding energy contributions), the binding energy of a single adsorbate can be written as

$$E_{\text{bind}} = -\frac{1}{N_{\text{ads}}} \left(E_{\text{ads@slab}}^{\text{total}} - E_{\text{slab}}^{\text{total}} - N_{\text{ads}} \cdot E_{\text{ads@gas}}^{\text{total}} \right) \qquad , \qquad (2.8)$$

where N_{ads} indicates the number of adsorbed atoms, $E_{\text{ads@slab}}^{\text{total}}$, $E_{\text{slab}}^{\text{total}}$, $E_{\text{ads@gas}}^{\text{total}}$ are the total energies of the slab with adsorbates, the clean slab and the adsorbates in the gas phase reference configuration. In this work, the gas phase reference is always the most stable gas phase configuration and therefore a negative binding energy indicates an exothermic adsorption process at $T = 0$ K.

2.2.4 Spurious Lateral Interactions in Simulations

At this point some remarks of caution are in place. In a lot of simulation computer codes and in particular in electronic structure codes for bulk systems, periodic boundary conditions are employed. In case of computer codes which are describing the electronic structure of solids the reason for the usage of periodic boundary conditions is simply originated in the

2.2. Quantities Provided by Simulations

fact that not only the geometric structure of a crystalline solid can be described by a crystal unit cell which is periodically repeated, but also the electronic structure of solids obey this periodicity. (Blochś theorem, see Eq. 3.32) Therefore, an elegant way to describe the electronic structure of a crystalline solid with the least computational effort will make use of this (for more details on periodic electronic structure calculations, see section 3.3). This specific way of implementation has a potential huge impact on the extraction of lateral interaction parameters. Therefore, in this section, these consequences are briefly illustrated.

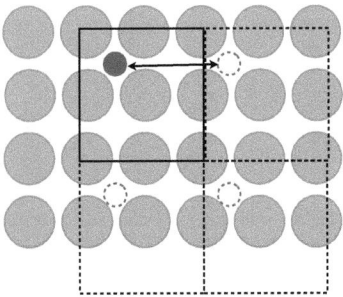

Figure 2.2: Spurious lateral interaction of a particle adsorbed on a surface with itself in a computer simulation employing periodic boundary conditions.

In a periodic code, the simulation cell is usually periodically repeated in *all three* dimensions. These repeated unit cells are schematically shown in figures 2.2, and 2.3. The red circle represents an adsorbed particle. Periodic repetitions are drawn with dashed lines in figure 2.2. Obviously, the adsorbed particle can in principle interact with its own periodic images in the surface plane. This will naturally influence the bonding energy, even in the case of a single adsorbate, as long as the size of the simulation cell is smaller in the xy plane than the decay length of the long range interactions. This is in particular important for the determination of the bonding energy of a single adsorbate which represents a reference energy for the determination of the amount of energy going into interactions.

Since the repetition of simulation cells is also done in z direction, metal surfaces are usually modelled in the so-called *supercell*-approach. This means that the simulation cell is composed of a bulk part, representing the truncated bulk and the surface and a vacuum region in z direction above the top layer. This vacuum region should basically represent the infinite vacuum area above a real surface, but is naturally limited by the bottom of the next periodic image in z direction. Therefore, to avoid a spurious coupling of the adsorbate with the bottom layer of the next simulation cell, one has to ensure that the vacuum is large enough to decouple the wavefunctions of each unit cell.

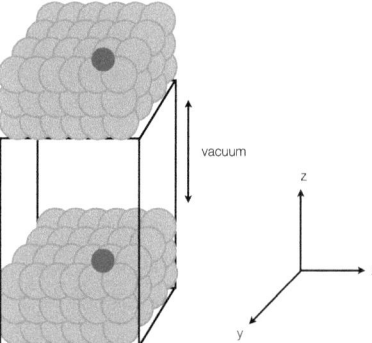

Figure 2.3: Since boundary conditions are used in all three space directions also images in z direction can couple with each other in the supercell approach and introduce artefacts into the simulations.

The formally correct way to overcome all these difficulties caused by the periodic images is, of course, to carefully check for convergence of the interesting parameters (the bonding energy of a single adsorbate, for example) with respect to the separation distances provided by a specific simulation cell setup.

Part II

Methods

Chapter 3
Methods - Electronic Structure Theory

*It is also a good rule
not to put too much confidence
in experimental results
until they have been confirmed by theory.*

— Sir Arthur Eddington

IN this study methods of electronic structure theory will mainly be used as a tool to generate highly accurate input data sets of total energies from first-principles. Based on these data sets different types of coarse grained models will be parameterized and used for larger scale simulations. This parameterization will furthermore allow the identification of the fundamental physical principles that are determining the parameterized interactions at a larger scale. The main objective for both descriptions is always the total energy of a certain configuration.

In this chapter, the fundamental aspects of the highest accuracy level — the electronic structure level — will be briefly presented. Starting from a many body description some fundamental aspects of wave function based electronic structure methods will be introduced. For a more thorough introduction to quantum chemistry, the reader may be referred to the textbook by Szabo and Ostlund [22]. As a next step, a quick overview about the formalism of density-functional theory (DFT) will be given as a different approach to tackle the total energy calculation from first-principles. For the specific introduction to density-functional theory the reader may be pointed to the books by Holthausen and Koch [23], Dreizler and Gross [24], or Parr and Yang [25].

In all first-principles calculations used in this work, exclusively the DFT method has been used. Because the systems under investigation are surface-adsorbate systems involving solids, one part of this chapter will present the DFT formalism for periodic systems. Finally, a very brief section will describe the usage of the electron density distribution information contained in the DFT results to simulate scanning tunneling microscopy (STM) images to directly compare to experimental data.

3.1 The Schrödinger Equation of a Many-Body System

For a material science context the most fundamental properties of complex systems are determined by the quantum mechanical behavior of their fundamental constituents, ions and electrons. This behavior can be described by the many-body wave function, Ψ, of the system[26]. Therefore, computational access to the detailed electronic structure described by the many-body wave function enables one to predict or evaluate the behavior or properties of a certain system. The internal groundstate energy is hereby the fundamental quantity that is of interest for the scope of the work presented in this thesis.

The total energy, E_{tot} of a many-body system that is described by the wave function, Ψ, is given as the expectation value of the Hamiltonian operator, $\hat{H}$, of this system. In Dirac's notation this writes:

$$E_{\text{tot}} = \langle \Psi | \hat{H} | \Psi \rangle \tag{3.1}$$

Given that the system under investigation comprises N electrons and M nuclei, solving Eq. 3.1 means solving a partial differential equation of $3(N + M)$ variables.

If we consider a system of M nuclei A of charge Z_A and N electrons (index i) without an external field, the Hamiltonian, in atomic units (i.e. $m_e = \hbar = e = 1$), is given by

$$\hat{H}_{\text{tot}} = \hat{T} + \hat{V} \quad , \tag{3.2}$$

where

$$\hat{T} = \hat{T}_e + \hat{T}_n = \underbrace{-\frac{1}{2}\sum_A^M \nabla_A^2}_{\text{for the nuclei}} - \underbrace{\frac{1}{2}\sum_i^N \nabla_i^2}_{\text{for the electrons}} \tag{3.3}$$

stands for the contribution of the kinetic energy of nuclei (n) and electrons (e). The potential energy operator for such a system $\hat{V}$ is written as

$$\hat{V} = \hat{V}_{\text{ne}} + \hat{V}_{\text{nn}} + \hat{V}_{\text{ee}} \quad = \quad \underbrace{-\sum_A^M \sum_i^N \frac{Z_A}{|\mathbf{R}_A - \mathbf{r}_i|}}_{\text{nuclear-electron (ne) attraction}}$$

$$+ \underbrace{\sum_A^M \sum_{B>A}^M \frac{Z_A Z_B}{|\mathbf{R}_A - \mathbf{R}_B|}}_{\text{nuclear-nuclear (nn) repulsion}} \tag{3.4}$$

$$- \underbrace{\sum_i^N \sum_{j>i}^N \frac{1}{|\mathbf{r}_i - \mathbf{r}_j|}}_{\text{electron-electron (ee) repulsion}} \quad .$$

Here, the potential energy contribution is written as the sum of attractive and repulsive interactions of nuclei and electrons, where $\mathbf{R}$ indicates positions of nuclei and $\mathbf{r}$ indicates positions of electrons.[I]

[I]Bold identifiers are vectors.

3.1. The Schrödinger Equation of a Many-Body System

The total Hamiltonian for the system can be written as the sum of all individual operators:

$$\hat{H}_{tot} = \hat{T}_e + \hat{T}_n + \hat{V}_{ne} + \hat{V}_{nn} + \hat{V}_{ee} \quad . \tag{3.5}$$

In a first step to simplify the mathematical complexity of this problem, one mostly well justified argument states that due to the huge difference in the resting mass between electrons and nuclei, electrons will almost instantaneously adapt to any change in position of the nuclei. Even for the worst case of the hydrogen atom, the ratio between the nucleus' mass and the mass of the electron is $\approx 1/1800$. Therefore, one can separate the problem into a purely electronical one and cover the (then approximately static) effects by the charged nuclei with the introduction of a static energy contribution.

By this so-called *Born-Oppenheimer* approximation [27] the dimensionality of the problem is reduced to the electronic degrees of freedom, therefore, resulting in $3N$ variables for which the Schrödinger equation needs to be solved. The resulting energy will be the electronic energy. In order to regain the total energy, the nuclear-nuclear repulsion terms must be added.

The Hamiltonian (BO Hamiltonian) takes then the following form in atomic units

$$\hat{H}_{BO} = -\frac{1}{2}\sum_i^N \nabla_i^2 - \sum_A^M \sum_i^N \frac{Z_A}{|\mathbf{R}_A - \mathbf{r}_i|} + \sum_i^N \sum_{j>i}^N \frac{1}{|\mathbf{r}_i - \mathbf{r}_j|} \quad . \tag{3.6}$$

For the evaluation of the ground state electronic energy, E_{BO}, the lowest set of eigenvalues of the spectrum of the BO Hamiltonian operator needs to be evaluated. Since electrons are *fermions* and thus must obey the *Pauli-Principle* [28,29], finding the groundstate energy equals the task of finding the energy minimum under arbitrary variations in a set of *anti-symmetric* wave functions, $\{\Psi\}$.

Consider an approximate wave function Ψ_{app} and the associated expectation value of the energy E_{app}. The *variational theorem* for bound operators states that the energy calculated from a trial Ψ_{app} is an upper limit to the true ground state energy. Minimisation with respect to all allowed antisymmetric N-electron wave functions, $\{\Psi_{app}\}$, will give the true ground state energy and the exact wave function. This minimization of the functional $E_{app}[\Psi_{app}]$ is known as the *Rayleigh-Ritz* minimal principle [30,31].

Therefore,

$$E_{app} = E_{0,BO} \iff \Psi_{app} = \Psi_0 \quad \text{i.e.} \quad E_{0,BO} = \min_{\Psi_{app}} E[\Psi_{app}] \quad . \tag{3.7}$$

This minimization with respect to all allowed electronic wave functions of the set $\{\Psi_{app}\}$ is still not practical. The entire field of so-called wave-function based quantum chemistry deals with the identification of useful approximate, but tractable forms for the many-body trial wave-function.

Hartree Wave-Function and Slater Determinant – The Basis for Hartree-Fock Methods

A very simple approach is to construct the molecular wavefunction as a combination of *independent* one–electron wave functions. This is the independent-particle, or Hartree-approximation [32,33], where the total wave function is built up as a product of orthonormal one–electron wave functions. This approximation assumes that each electron moves independently in its own orbital, only influenced by the *average* field generated by all the other electrons and explicit electron interaction is neglected.

This simple wave-function is an invalid one, because it is not antisymmetric with respect to electron interchange and therefore violates the Pauli-Principle, which imposes an essential principle for the stability of matter [34,35].

Linear algebra yields that a valid wave function, which changes sign upon particle exchange, i.e. is anti-symmetric, can be constructed by the so-called *Slater determinant* [36] of one–electron wave functions.

A Slater-Determinant forms the basis for the famous Hartree-Fock (HF)calculations [37], in which the starting trial wave function is minimized iteratively under the constraint that all one–electron wave functions remain orthonormal. By including the anti-symmetric permutations, the contributions of electron *exchange* interactions are covered within Hartree-Fock calculations.

However, within this scheme explicit electron-electron interaction is neglected and replaced by an averaged contribution. Electrons of the same spin are partially but not completely, correlated. No attempt is made to include electron-correlation of opposite-spin electrons.

Because of this neglect, the energy of HF calculations, E_{HF} is always above the exact non-relativistic value, E_{exact}, the difference is defined as the *correlation energy*, $E_{corr} = E_{exact} - E_{HF}$. This insufficient treatment of the electron correlation is the main deficiency of HF theory. Usually, in a sufficiently large basis set HF is able to account for as much as 99% of the exact groundstate energy, but the remaining 1% is often very important for describing chemical phenomena [38].

Post Hartree-Fock methods try to overcome this deficiency by accounting for electron correlation in different ways. Some examples for those approaches are second or higher order Møller-Plesset (MP2, MPn) perturbation theory [39], configuration interaction (CI), or coupled cluster (CC) approaches [40]. The major disadvantage of those methods is their very unfavorable scaling, so that even today they are not generally applicable to larger molecules with many electrons. The most often used method in the quantum chemical community, however, is MP2 as for this the scaling behavior (N^5) is still manageable for most systems and a smart implementation can lower its scaling [38].

3.2 Density-Functional Theory

A different starting point for the evaluation of the total energy is taken by density-functional theory. Instead of trying to construct the best all-electron wavefunction, Ψ, (which depends, as already stated on $3N$ variables — $4N$ if spin is included) the idea is to rewrite the Schrödinger equation in terms of the natural and physical low-complexity quantity, the electron density, $n(\mathbf{r})$.

This reformulation leads to a decrease in the dimensionality of the problem, leaving 3 variables, *independently on the number of electrons*. While the complexity of the wave function increases with the number of electrons, the electron density has the same number of variables, independently of the system size.[38] If there exists a bijective mapping between the system's Hamiltonian and the electron density it means that *every* observable of the ground-state system can be calculated from the density *alone* and it does not matter how that density then is obtained.

The electron density is defined as the the number of electrons per unit volume in a given state. For a given state with wave function Ψ the correlated electron density is given by

$$n(\mathbf{r}) = N \int \Psi^* \cdot \Psi \, ds_1 d\mathbf{x}_2 d\mathbf{x}_3 \ldots d\mathbf{x}_N \tag{3.8}$$

with the normalization criterion

$$\int n(\mathbf{r}) d\mathbf{r} = N \quad . \tag{3.9}$$

The DFT total energy , E_{DFT}, is then expressed as a *functional* of the electron density of the system, which in turn is determined by the wave function. All terms of the BO Hamiltonian are now expressed as functionals with the electron density as the variable.

3.2.1 Thomas-Fermi Theory

Already as early as in the 1920s in the model of Thomas and Fermi [41,42] statistical considerations were used to approximate the distribution of the electrons in an atom and derive a general energy functional only depending on the electron density. Based on the assumptions of (i) treating the electrons as independent particles, (ii) reducing the electron-electron interaction to the Coulomb electrostatic energy and (iii) using a *local density approximation* for the kinetic energy a functional for the total energy was derived.

The energy functional in Thomas-Fermi theory, $E^{\text{TF}}[n(\mathbf{r})]$, is composed of three terms

$$E^{\text{TF}}[n(\mathbf{r})] = A_k \int n(\mathbf{r})^{5/3} d\mathbf{r} + \int n(\mathbf{r}) v_{\text{ext}}(\mathbf{r}) d\mathbf{r} + \frac{1}{2} \iint \frac{n(\mathbf{r})n(\mathbf{r}')}{|\mathbf{r}-\mathbf{r}'|} d\mathbf{r} d\mathbf{r}' \tag{3.10}$$

in which the *external potential* v_{ext} is given by the static Coloumb potential arising from the nuclei

$$v_{\text{ext}}(\mathbf{r}) = -\sum_{A}^{M} \frac{Z_A}{|\mathbf{r}-\mathbf{R}_A|} \quad . \tag{3.11}$$

The first term of equation 3.10 is the electronic kinetic energy associated with a system of non-interacting electrons in a homogenous electron gas. It is obtained by integrating the kinetic energy density of a homogeneous electron gas $t_0[n(\mathbf{r})]$[43]:

$$T^{\text{TF}}[n(\mathbf{r})] = \int t_0[n(\mathbf{r})]\mathrm{d}\mathbf{r} \quad , \tag{3.12}$$

where $t_0[n(\mathbf{r})]$ is obtained by summing all the free-electron energy states $\varepsilon = k^2/2$ up to the Fermi wavevector $\mathbf{k}_F = [3\pi^2 n(\mathbf{r})]^{1/3}$, which then leads to

$$t_0[n(\mathbf{r})] = \frac{2}{(2\pi)^3} \int \frac{\mathbf{k}^2}{2} n_{\mathbf{k}} \mathrm{d}\mathbf{k} = \frac{1}{2\pi^2} \int_0^{\mathbf{k}_F} \mathbf{k}^4 \mathrm{d}\mathbf{k} \tag{3.13}$$

and finally determines the coefficient A_k to be $A_k = \frac{3}{10}(3\pi^2)^{2/3} = 2.871$. Finally, the last term in the energy functional describes the classical Coulomb repulsion between the electrons.

To obtain the groundstate density and by that the goundstate energy the functional is minimized using the technique of Lagrange multipliers — denoted μ here — to impose the constraint of equation 3.9. This yields the so-called Thomas-Fermi equation

$$\frac{5}{3} A_k n(\mathbf{r})^{2/3} + v_{\text{ext}}(\mathbf{r}) + \int \frac{n(\mathbf{r}')}{|\mathbf{r}-\mathbf{r}'|} \mathrm{d}\mathbf{r}' - \mu = 0 \tag{3.14}$$

Thomas-Fermi theory fails for real systems and does not predict bonding, mainly due to the bad description of the kinetic energy contribution of the electrons.[44–46] An additional shortcoming of Thomas-Fermi theory is that it neglects the electron exchange contribution in the original formulation, like Hartree theory does.

Shortly after the publication of Thomas-Fermi theory, Dirac[47] developed an approximation to include the exchange interaction contributions into it based on the homogeneous electron gas. The exchange contribution can be simply given as a local function of the density:

$$E_x[n(\mathbf{r})] = -\frac{3}{4}\left(\frac{3}{\pi}\right)^{1/3} \int n(\mathbf{r})^{4/3} \mathrm{d}\mathbf{r} \quad , \tag{3.15}$$

which can be also expressed in terms of the exchange energy density $\varepsilon_x[n(\mathbf{r})]$ as,

$$E_x[n(\mathbf{r})] = \int n(\mathbf{r})\varepsilon_x[n(\mathbf{r})]\mathrm{d}\mathbf{r} \quad , \tag{3.16}$$

with the possibility to express $\varepsilon_x[n(\mathbf{r})]$ then simply in terms of the Seitz radius — the radius of a sphere that contains the charge of one electron — r_s [48]:

$$\varepsilon_x[n(\mathbf{r})] = -\frac{3}{4}^{1/3} \frac{1}{r_s} \approx -\frac{0.4582}{r_s} \quad . \tag{3.17}$$

Incorporation of the exchange contribution into the Thomas-Fermi functional is simply done by adding the term of equation 3.16 to the functional of the kinetic energy contribution, equation 3.12, and account for it with a term of $\frac{4}{3}\varepsilon[n(\mathbf{r})]$ in the resulting Euler-Lagrange form in the minimization. However, also this inclusion does not lead to more meaningful predictions for real systems.

3.2.2 The Hohenberg-Kohn Theorems

The foundation of modern density functional theory was laid by the famous paper of Hohenberg and Kohn from 1964[49]. In this paper, two theorems were now strictly proven that allowed this systematic connection between the ground state electronic density distribution of a system and its ground state energy.

The first theorem states that there is a one-to-one correspondence between the electron density of a system and its energy and vice versa. This theorem can be proven with the variational principle.

The second Hohenberg-Kohn theorem adopts the variational principle into density functional theory. For a trial density $\tilde{n}(\mathbf{r})$, such that $\tilde{n}(\mathbf{r}) \geq 0$ and $\int \tilde{n}(\mathbf{r}) d\mathbf{r} = N$, an energy $E[\tilde{n}(\mathbf{r})] \geq E^0$ arises. This implies, an incorrect density will yield a energy which is higher than the exact ground state energy for every trial density supplied by a trial wave-function.

3.2.3 Kohn-Sham Density-Functional Theory

The foundation for the practical use of DFT in computational chemistry was the introduction of orbitals by *Kohn* and *Sham*[50]. With this introduction, the treatment of the main problem in Thomas-Fermi Theory can be significantly improved. The basic idea is to minimize the error in the kinetic energy term by splitting the kinetic energy into two parts, one of which can be calculated exactly, and a small correction term.

Consider a system of *non-interacting* electrons. For this system, the kinetic energy is described exactly with the functional of the Hartree-Fock-Slater theory, that is

$$T_s = \sum_i^N \langle \chi_i | \hat{T} | \chi_i \rangle$$
$$= \sum_i^N \langle \chi_i | -\frac{1}{2} \nabla^2 | \chi_i \rangle \quad . \tag{3.18}$$

The electron density for this system is given exactly by

$$n(\mathbf{r}) = \sum_i^N |\chi_i(\mathbf{r})|^2 \tag{3.19}$$

and the total energy is given by

$$E[n(\mathbf{r})] = T_s[n(\mathbf{r})] + V_{ne}[n(\mathbf{r})] + J[n(\mathbf{r})] \quad . \tag{3.20}$$

By this splitting the major contribution to the kinetic energy functional of the interacting system is described correctly. Kohn and Sham reformulated the interacting problem so that the difference between the exact kinetic energy and the one calculated for non-interacting orbitals is absorbed into an *exchange-correlation* term, including also the exchange part $K[n(\mathbf{r})]$. In general, a KS-DFT energy expression can be written as

$$E_{\text{DFT}} = T_s[n(\mathbf{r})] + V_{ne}[n(\mathbf{r})] + J[n(\mathbf{r})] + E_{xc}[n(\mathbf{r})] \quad . \tag{3.21}$$

Then, the exchange-correlation term, $E_{xc}[n(\mathbf{r})]$ is defined by this equation (3.21). This, E_{xc}, exchange correlation energy can not completely be compared with the terms of Hartree-Fock theory, because the definitions are different. The E_{xc} depends uniquely on the total electronic density distribution. Through its functional derivative

$$v_{xc}(\mathbf{r}) = \frac{\delta E_{xc}[n_0(\mathbf{r})]}{\delta n_0(\mathbf{r})} \quad , \tag{3.22}$$

it determines an additional potential to be taken into account in the search for the ground state electron density. It has been shown that this property of the so-called v-*representability* is fulfilled by any physical charge density on a lattice [51]. $V_{ne}[n(\mathbf{r})]$ can be written in terms of the external potential, $v(\mathbf{r})_{ext}$, and the electron density as

$$V_{ne}[n(\mathbf{r})] = \int n(\mathbf{r}) v_{ext}(\mathbf{r}) \, d\mathbf{r} \quad . \tag{3.23}$$

Now one has to iteratively find the set of orbitals, yielding the density, which minimizes the energy to a certain threshold. Again the method of choice is the use of Lagrange multipliers. The equations are arranged with a one-electron operator, similar to the Fock-operator and the set of coefficients making the eigenvalue equations stationary with respect to the energy must be found. This yields the Kohn-Sham orbitals of the system, which will describe the electron density correctly and, as a result, yield the correct energy.

The eigenvalue equations are known as *Kohn-Sham equations*.

$$\left[-\frac{\nabla^2}{2} + v_{ext}(\mathbf{r}) + \int \frac{n(\mathbf{r})}{|\mathbf{r} - \mathbf{r}'|} \, d\mathbf{r}^3 + v_{xc}(\mathbf{r}) - \varepsilon_j \right] \phi_j(\mathbf{r}) = 0 \quad , \tag{3.24}$$

where ϕ_j describes the *Kohn-Sham* (KS) orbitals.

Because the KS-orbitals only have to decribe the correct density as opposed to the HF-orbitals, which have to describe the complete electronic properties, the requirements for their accuracy are not as high as in Hartree-Fock, which allows the use of lower order basis sets and therefore in this treatment some computational cost can be saved. However, orbitals do introduce a more extensive size-dependence and therfore a higher computational cost into density-functional theory again.

The problem which still remains is to find the correct exchange-correlation functional. There is no systematic way to find such a functional. If the exact functional was known, DFT would provide the exact groundstate energy. Still, there are a number of well known approximations which yield results in the range of high order wave mechanics approaches.

3.2.4 DFT Exchange-Correlation Functionals

As already stated, the major deficiency in Hartree theory is the fact that none of the quantum mechanical electron-electron interactions are taken into account in the construction of the system's wave-function, therefore neglecting all influences caused by exchange and correlation effects. Thus in Hartree theory we have $E_{xc}[n] = 0$. Now in DFT the purpose of the exchange and correlation (xc) functional, $E_{xc}[n]$, is to cover these effects and also include the contributions to the kinetic energy term by the interaction of the electrons.

3.2. Density-Functional Theory

The Local Density Approximation (LDA): To improve the treatment of exchange and correlation effects, a starting point is presented by the homogenous electron gas, for which an expression for the exchange correlation term is given by (neglecting the effects of spin-polarization for brevity)

$$E_{xc}^{\text{LDA}}[n] = \int n(\mathbf{r}) \varepsilon_{xc}^{\text{HEG}}[n(\mathbf{r})] \, d^3\mathbf{r} \quad (3.25)$$

where $\varepsilon_{xc}^{\text{HEG}} = \varepsilon_{x}^{\text{HEG}} + \varepsilon_{c}^{\text{HEG}}$ is the xc-energy of a homogeneous electron gas (HEG). The LDA is strictly local in space, and is exact in the limit of slowly varying densities (albeit in practice limited by the accuracy with which $\varepsilon_{xc}^{\text{HEG}}$ can be determined). The exchange part can be computed analytically, see equation 3.17[47].

The numerical value of the correlation part, $\varepsilon_{c}^{\text{HEG}}[n(\mathbf{r})]$ cannot be expressed in an explicit form. Expressions for the low-density limit[52,53] and the high-density limit[54,55] are known, whereas for intermediate densities only values from numerical quantum Monte-Carlo calculations[56] are known to high accuracy. All correlation parameterizations used today[57–59] are suitable numerical interpolations of these numerical values and the exact limits.

The LDA leads to useful results for most applications. Experience has shown that LDA gives ionisation energies of atoms and dissociation energies of molecules with a fair accuracy, typically of 10–20%. Bond lengths and thus the geometry of molecules have typically accuracies of ≈ 1% for main-group molecules. The results are not so good for transition metals. The solution of the Kohn-Sham equations in the LDA is only minimally more difficult than the solution of the Hartree equation and very much easier than the solution of the Hartree-Fock equations. The accuracy for the exchange part is about 10%, while the much smaller correlation part is generally overestimated with by factor of two. The two errors typically cancel partially. However, one has to keep in mind that this error cancellation is a mathematical coincidence and has not necessarily physical meaning.

Generalized Gradient Approximation (GGA) Since the assumption of a uniformly distributed electron density, which only varies slowly in space, is a rather strong one in particular for molecular systems a straightforward extension to the local density approximation is to take into account the gradient of the density in the calculation of the xc contributions. The next rung of xc-functionals used in modern DFT calculations is represented by the so-called generalized gradient approximation (GGA)

$$E_{xc}^{\text{GGA}}[n] = E_{xc}^{\text{LDA}}[n] + \int \Delta \varepsilon_{xc}[n(\mathbf{r}), |\nabla n(\mathbf{r})|] d^3\mathbf{r} \quad (3.26)$$

where $\Delta \varepsilon_{xc}[n(\mathbf{r}), |\nabla n(\mathbf{r})|]$ is an xc-energy gradient correction dependent on the local density $n(\mathbf{r})$, and the local reduced density gradient $\nabla n(\mathbf{r})$. Thus, the GGA is semi-local, rather than strictly local.

In the energy calculations of the level of density-functional theory, the GGA xc-functional representation as parametrized by Perdew, Burke and Ernzerhof (PBE)[60,61] has been exclusively used.

3.3 Application to Periodic Systems, Plane Waves and Pseudopotentials

For systems of an almost homogeneous electron gas the natural choice for basis functions are plane waves. Solids, in particular metals, can be seen as representations of such a periodic systems of a homogeneous electron gas with the ionic cores representing small pertubations of the homogenity. Therefore, the Kohn-Sham orbitals for a uniform electron gas are exactly given by plane waves,

$$\phi_{\mathbf{k}}(\mathbf{r}) = -1/2\Omega_c \exp\left[i\mathbf{k}\cdot\mathbf{r}\right] \qquad (3.27)$$

where Ω_c is the volume of the box as defined below, equation 3.29.

This chapter mainly represents a short summary of the very nice and educational review article of Meyer[62] on this topic.

3.3.1 Periodicity in Three Dimensions

The effective potential (as well as the electron density) is a periodic function with the periodicity of the lattice of the underlying periodic structure

$$V_{\text{eff}}(\mathbf{r}+\mathbf{T}) = V_{\text{eff}}(\mathbf{r}) \qquad , \qquad (3.28)$$

for any lattice vector $\mathbf{T}$, where $\mathbf{T} = N_x \mathbf{a}_x + N_y \mathbf{a}_y + N_z \mathbf{a}_z$.

The periodic unit cell is given here by the three primitive lattice vectors which define a periodically repeated volume element in real space by

$$\Omega_c = \mathbf{a}_i \cdot \left(\mathbf{a}_j \times \mathbf{a}_k\right) \qquad . \qquad (3.29)$$

(in which i, j, k are any arbitrary permutation of (x, y, z)).

Therefore, V_{eff} can be expanded in a Fourier series

$$V_{\text{eff}}(\mathbf{r}) = \sum_{\mathbf{G}} V_{\text{eff}}(\mathbf{G}) \exp\left[i\mathbf{G}\mathbf{r}\right] \qquad , \text{with} \qquad (3.30)$$

$$V_{\text{eff}}(\mathbf{G}) = \frac{1}{\Omega_c} \int_{\Omega_c} V_{\text{eff}}(\mathbf{r}) \exp\left[-i\mathbf{G}\mathbf{r}\right] d^3\mathbf{r} \qquad (3.31)$$

The sum runs over all vectors $\mathbf{G}$ which fulfill the condition $\mathbf{G}\cdot\mathbf{T} = 2\pi M$, with $M \in \mathcal{Z}$. Those vectors form the so-called *reciprocal lattice*, the primitive vectors $\mathbf{b}_x, \mathbf{b}_y, \mathbf{b}_z$ of which are given by the relation: $\mathbf{a}_i \cdot \mathbf{b}_j = 2\pi \delta_{ij} \qquad (i,j = x, y, z)$.

3.3.2 Bloch's Theorem and Expansion in Plane Waves

Solving now a single-particle Schrödinger equation for a periodic structure can be simplified dramatically by employing the famous *Bloch theorem*[63]. One formulation of this theorem states that all eigenfunctions $\psi_{\mathbf{k}_j}$ of a single-particle Schrödinger equation with a periodic

3.3. Application to Periodic Systems, Plane Waves and Pseudopotentials

potential can be written as a periodic function $u_{\mathbf{k}_j}$ modulated by a plane-wave with a wave vector $\mathbf{k}$:

$$\psi_{\mathbf{k}_j}(\mathbf{r}) = \exp[i\mathbf{k}\mathbf{r}] u_{\mathbf{k}_j}(\mathbf{r}) \quad (3.32)$$

This allows in the process of calculation to calculate only the eigenfunctions within one unit cell and the form of the eigenfunctions in all other unit cells is determined by making use of equation 3.32. Therefore, it is convenient to define that all eigenfunctions are normalized with respect to the unit cell. This is

$$\int_{\Omega_c} \psi_{\mathbf{k}_j}(\mathbf{r})^2 \, d^3\mathbf{r} = 1 \quad (3.33)$$

Since the functions $u_{\mathbf{k}_j}$ are periodic, they can be expanded in a set of plane waves. Taking into account the exponentiell prefactor, one gets:

$$\psi_{\mathbf{k}_j}(\mathbf{r}) = \sum_{\mathbf{G}} c_{\mathbf{G}}^{\mathbf{k}_j} \exp[i(\mathbf{k}+\mathbf{G})\mathbf{r}] \quad . \quad (3.34)$$

Here j is a band index and $c_{\mathbf{G}}^{\mathbf{k}_j}$ are the expansion coefficients of the plane wave expansion.

One can now write the one-electron Kohn-Sham equations in the notation of the Bloch-states, represented by $\mathbf{k}$.

$$\left(-\frac{\hbar^2}{2m}\Delta + V_{\text{eff}}(\mathbf{r})\right)\psi_{\mathbf{k}_j}(\mathbf{r}) = \varepsilon_{\mathbf{k}_j} \psi_{\mathbf{k}_j}(\mathbf{r}) \quad , \quad (3.35)$$

where $V_{\text{eff}}(\mathbf{r}) = V_{\text{ext}}(\mathbf{r}) + V_H[n(\mathbf{r})] + V_{xc}[n(\mathbf{r})]$ and the density $n(\mathbf{r})$ is calculated using the eigenfuntions.

3.3.3 Some Technical Points about Solving the Equations Using Plane Waves

k-point Sampling

By using Bloch's theorem the problem of calculating an infinite number of electronic states in an infinite space has been mapped to one of calculating a finite number of eigenstates at an infinite number of k-points which are extended over a single unit cell. In principle, still an infinite number of calculations would be necessary for the different k-points. However, by arguing that the electronic wave functions at different k-points which are close to each other will be very similar, one opens the possibility to describe all the wavefunctions of a region in k-space just by one wave-function at a single selected k-point. One thus follows the approach to define a regular mesh of N_{kpt} k-points and replaces the integrals over the Brillouin zone by a finite and discrete sum over the chosen k-point mesh:

$$\frac{\Omega_c}{(2\pi)^3}\int_{\text{BZ}} \ldots \to \frac{1}{N_{\text{kpt}}}\sum_{\mathbf{k}} f_{\mathbf{k}_j} \ldots \quad (3.36)$$

The $f_{\mathbf{k}_j}$ are occupation numbers which are either one or zero, in principle. There are several different schemes to construct k-meshes around in the literature[64–68]. The one used in all periodic DFT calculations in this thesis is the one by Monkhorst and Pack[65].

This approach allows the calculation of the total energy of a solid by evaluation of the electronic states only at a finite number of k-points. The error which is introduced by this approximation can be reduced systematically by increasing the density of the employed k-mesh. Using denser and denser k-meshes a convergence test can be performed to determine the density of the k-point mesh for a desired accuracy of the quantity of interest.

Since the k-points are distributed in reciprocal space and the volume of the Brillouin zone becomes smaller and smaller with an increasing size of the real space supercell of the calculation, less and less k-points are usually needed with increasing supercell size. From a certain supercell size on it is often justified to evaluate the electronic wave-function at just a single k-point, which is usually the origin, the so-called Γ-point.

To accelerate convergence for metallic systems the introduction of fractional occupational numbers by certain schemes[67,69–71] is often used. However, if used also the influence of those smearing width value should to be studied for convergence behavior.

Maximum G Vector – Cut-Off Energy

The matrix eigenvalue equation of the Kohn-Sham equations within the plane wave representation are given by:

$$\sum_{\mathbf{G}} \left(\frac{\hbar^2}{2m} \|\mathbf{k}+\mathbf{G}\|^2 \delta_{\mathbf{G'G}} + V_{\text{eff}}(\mathbf{G'}-\mathbf{G}) \right) c_{\mathbf{G}}^{\mathbf{k}_j} = \varepsilon_{\mathbf{k}_j} c_{\mathbf{G}}^{\mathbf{k}_j} \quad . \tag{3.37}$$

Similar to the fact that in principle one should calculate the eigenstates at an infinite number of k-points and the infinite integral is replaced by a discrete summation over special k-points, one also truncates the Fourier expansion after some final lattice vector $\mathbf{G}$.

The sum $(\mathbf{G}+\mathbf{k})$ represents the plane wave vector and the criterion for the truncation is defined by a maximal plane wave kinetic energy, E_{cut}:

$$\frac{\hbar^2}{2m} \|\mathbf{k}+\mathbf{G}\|^2 \leq E_{\text{cut}} \tag{3.38}$$

Also here systematic convergence tests are made possible by the fact that higher plane wave cutoff energies systematically improve the accuracy of the energy evaluation. The electron density in Fourier representation is given by

$$n(\mathbf{G}) = \frac{2}{N_{\text{kpt}}} \sum_{\mathbf{k}_j} f_{\mathbf{k}_j} \sum_{\mathbf{G'}} \left(c_{\mathbf{G'}-\mathbf{G}}^{\mathbf{k}_j} \right)^* c_{\mathbf{G'}}^{\mathbf{k}_j} \quad . \tag{3.39}$$

The evaluation of the Hartree potential is particularly convenient in Fourier space

$$V_{\text{H}}(\mathbf{G}) = 4\pi e^2 \frac{n(\mathbf{G})}{|\mathbf{G}|^2} \tag{3.40}$$

Due to the Fourier expansion only being performed up to a maximal wave-vector both quantities have non-vanishing Fourier components only up to twice the length of this cut off wave vector, therefore the expansions are finite and easy to evaluate.

3.3.4 Pseudopotentials

So far plane waves seem to be the ideal basis for electronic structure calculations of solids. Besides their mathematical quite simple representation and their ability to enable elegant evaluations of certain quantities in reciprocal space by means of fast-fourier transformations, they are systematically improvable up to the desired accuracy.

One major disadvantage of a pure plane wave representation of the electronic wave function lies in the fact that a proper description of the strong oscillating nature of the electronic wavefunction close to the atom cores requires a quite large number of superpositioned plane waves, which leads to the necessity to use high cutoff energies.

For the proper description of many physical problems, however, the core electrons are not important. Bonding and other properties of interest are almost purely determined by the distribution and interaction of the valence electrons. The core electrons are strongly localized at the nuclei and their wave functions overlap only very little. These considerations are the basic justification for the usage of the so-called *frozen-core approximation* in electronic structure theory. The charge of the pure nuclei is replaced by an effective core charge which is constructed from the core potential and the Hartree potential of the core electrons. Orthogonality is ensured by the employed pseudoization mathematical procedure. This approach has significant advantages. The number of electrons which needs to be described by the wave-function is dramatically reduced, which significantly reduces the computational cost of such a calculation.

An important thing to note is, however, that by construction the pseudopotential shifts the true total DFT potential by an unphysical constant. For this reason only differences between pseudo total energies or their variation with physical variables correspond to physical quantities. The total energy of a pseudopotential calculation itself does not.

This very short overview is not intended to go into the details of the construction of pseudopotentials. However, it needs to be stated that for matters of numerical efficiency in all electronic structure calculations presented in this work *ultasoft Vanderbilt*-type pseudopotentials[72,73] within the plane-wave code CASTEP[74] have been used for density-functional based first-principles energy evaluations.

3.4 STM Simulations Using the Tersoff-Hamann Model

A quite convenient connection between the results of electronic structure calculations to experimental data can be established by *scanning tunneling microscopy* (STM) experiments. In a STM experiment, the local electronic structure is probed with the STM tip.

The general theory of this experimental technique represents a quite complicated many-body problem[75], since there is not necessarily a simple interaction between the STM tip and the sample electronic structure. However, even the simplified picture of the so-called Tersoff-Haman model[76,77] can argued to be justified at large tip-sample distances and usually produced reasonably well agreement between experiments and theory.

In this picture, first-order perturbation theory is applied for the tip-sample coupling, which leads to an approximate expression for the tunnel current between the two interacting electrodes:

$$I = \frac{2\pi e}{\hbar} \sum_{\mu\nu} f(E_\mu)[1 - f(E_n u)] |M_{\mu\nu}|^2 \delta(E_\mu - (E_\nu + eV)) \quad . \quad (3.41)$$

In this expression $f(E)$ represents the Fermi-function, V is the applied tunneling voltage and $M_{\mu\nu}$ is the tunneling matrix element between the tip states ψ_μ and the surface states ψ_ν. In the limit of small voltage and temperature, Eq. (3.41) can be simplified under the assumptions[II] of the Tersoff Haman model to:[78]

$$I \propto \sum_\nu |\psi(\mathbf{r}_t)|^2 \delta(E_\mu - E_\nu) = \int_{E_F}^{E_F + \Delta E} dE \rho(\mathbf{r}_t, E) \quad . \quad (3.42)$$

This expression describes the tunneling current as proportional to the local density of states of the surface at the tip position $\rho(\mathbf{r}_t, E)$ in the small energy window around the Fermi level, E_F, and the applied tunneling voltage $\Delta E = V \cdot e$. The local density of states represents the eigenstates of the orbitals at a specific energy value. In a density-functional theory electronic structure code, the calculation of this quantity is carried out by a summation about the properly weighted Kohn-Sham orbitals.

The resulting electron density can then be plotted according to the two fundamental operation modes of a STM measurement.

- In the *constant-current* mode, the xyz coordinates of a specific isovalue of the density, representing a certain tunnel current, are plotted, usually xy resolved and colored according to the z value.

- In the less common *constant-height* mode, one plots the xy resolved isolvalue of the electronic density at a given z coordinate.

All STM simulations performed in this work are *constant-current* simulations, which are used to resolve the geometric structure of the electronic density.

In addition, it is useful to clarify the influence of the sign of the bias voltage on the sampling of orbitals. A negative potential on the sample probes occupied sample states, whereas a positive sample potential measures unoccupied states of the sample.

[II]Those are very briefly, (i) the tip can be represented by a spherical symmetric s-like orbital, (ii) there is no interaction between tip and sample, (iii) the current is imply proportional to the local density of states of the surface at the lateral position of the tip

Chapter 4
Bridging Scales: Connection of Microscale Energetics and Mesoscale Functionality

Your problem is to bridge the gap which exists between where you are now and the goal you intend to reach.

Earl Nightingale

ALTHOUGH one could in principle calculate or derive all properties of interest from electronic structure calculations, these calculations are in practice computationally so demanding that one faces certain scale limitations. Even with the approximations and methods to make those calculations feasible for smaller systems that have been introduced in the chapters before, the scales that are reachable are far too small to approach real-world dimensions. However, the detailed electronic structure information about a certain compound is not necessarily needed for the determination of certain properties, in particular at bigger scales. The quantity that is determining most of the material specific properties is obviously the energetics.

According to the property of interest, a certain time and length scale of the system under investigation must be described within the theoretical treatment. Therefore, it is essential to pick the right representation according to the quantity one is interested in. If this is now only the system's energetics, a proper description of these can prove itself valid and sufficient to derive already a large set of information about the material specific functionalities.

If one is interested in more than just a single ground state energy of one specific configuration of a system, i.e. in the state of the system at finite temperatures or at a mesoscopic scale, ensemble statistics naturally have to come into play. The connection between this mesoscopic regime and the microscopic regime is established by the partition function of the system. An exact mesoscopic treatment based on a microscopic picture could therefore be established by a calculation of the full partition function of the system in which each microscopic state would be treated on the level of electronic structure theory[79]. This approach, although formally correct, is computationally not yet feasible for practical systems like the

32   Chapter 4. Bridging Scales

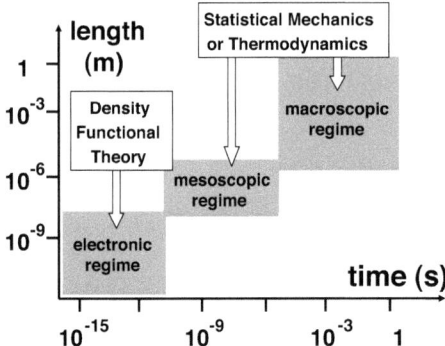

Figure 4.1: The different simulation methods can be grouped in a length and time scale representation. The electronic regime of a material, which determines the behavior and properties of it, is described by electronic structure methods, such as DFT. However, the macroscopic functionalities of systems are determined by the statistical interplay of the microscopic processes, which are determined by the principles of statistical mechanics and thermodynamics.
Corresponding to these scales are also the computational cost and the accuracy connected to the different regimes.

ones addressed in this thesis. The required number of *ab initio*/first-principles energy evaluations is too immense to be tackled with nowadays computer power. This problem motivates the search for approaches that do not rely on electronic structure calculations to calculate energies of various configurations.

As illustrated in Figure 4.1 different theories are necessary for the appropriate description of the governing mechanisms at the respective scale. In their own regimes, different theories are well established and properly founded. What is still missing nowadays, is a clean and proper connection of certain theories to cover several length and/or time scales within one methodological framework. The work of this thesis can be placed into the attempts to establish and investigate a linkage between energetics from the electronic regime (for which the methods used have been described in the chapter before) to the mesoscale of system's of interacting adsorbates.

The following two chapters (5, 6) will now introduce the methods used and/or developed within this thesis to establish this connection. A special focus will be placed in the investigation of how these mesoscale model descriptions are able to grasp the various kinds of governing interactions between particles adsorbed at a surface. Within the introduction of these methods, some emphasis will be laid on a proper description of how the large scale models are derived or connected to the underlying first-principles electronic structure calculations. These connections will be explained in detail in the corresponding chapter for each method separately. Finally, a chapter (7) will present the different techniques used in the search for the «best» structure/configuration. It will be distinguished between local and global search methods and the concepts of the employed method of each field that have been used in this thesis will be briefly described.

Chapter 5
Methods - Cluster Expansion

FOR a system of adsorbates on a surface in which the adsorbates are located at usually only a very small number of distinct adsorption sites of the surface facet, one quite elegant approach is founded on the idea of expressing the total energy of the given configuration by an Ising-like Hamiltonian. This method is called *Cluster Expansion(CE)* or *Lattice-Gas Hamiltonian (LGH)* approach. The basic idea originates from the description of ordering phenomena in spin systems [80,81], is common in the alloy community [82,83] but has also been expanded to surfaces [84–89].

5.1 Calculating the Total Energy in the Cluster Expansion Formalism

The Hamiltonian of the total system is written as an expansion of the configurational system's energy in terms of 0th order interactions (which are the contributions by direct bonding to the substrate), 1st order interactions (interacting pairs of adsorbed molecules), 2nd order interactions (three adsorbed molecules interaction), etc.... Formally, this can be expressed as:

$$E^{\text{total}}(\text{configuration}) = \sum_i^{N_{\text{atoms}}} E^{\text{onsite}}_{\text{bond}} + \sum_{(i,j)} V^{\text{pair}}(i,j) + \sum_{(i,j,k)} V^{\text{trio}}(i,j,k) + \ldots \quad , \quad (5.1)$$

in which $E^{\text{onsite}}_{\text{bond}}$ represents the energy contribution to the configurational energy gained by the pure binding to the surface at the specific adsorption position of the N_{atoms} particles. $V^{\text{pair}}(i,j)$ and $V^{\text{trio}}(i,j,k)$ represent the energy contributions by the adsorbate–adsorbate interactions of two neighboring particles (*pair*) or three neighboring particles (*trio*), respectively.

The corresponding summations are running over all pairs of a given kind, e.g. nearest-neighbor pairs or next-nearest-neighbor pairs, ... and trios in the first neighbor shell of atoms ... etc. If one would proceed with this expansion to infinite higher order many-body interactions, this would in principle be an exact approach to the total energy[^1] for an arbitrary configuration[80]. However, in practice one is forced to truncate the expansion after some higher order interaction. This is usually well justified, since the interaction strength usually decays very fast with large distances between the interacting adsorbates.

The determination of the energy contributions represented by the different interactions can then be performed in a fitting procedure. This can be either done by fitting to a variety of experimental data such as heats of adsorption, phase diagrams or thermal desorption data. By such an approach, one can obviously gain useful insight about the interaction parameters and their nature. However, one clearly lacks a *microscopic foundation* for the resulting values for the interaction parameters and will most likely not necessarily cover the underlying physics on the microscopic scale. Hence there will be most likely an unsatisfying general predictive power of the resulting CE Hamiltonian. Furthermore, a clear connection between mesoscopic scale and microscopic scale is obviously missing.

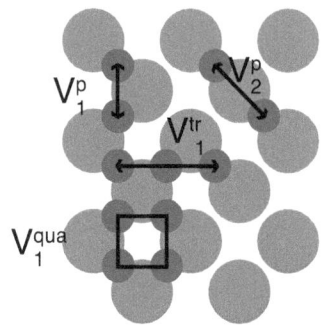

Figure 5.1: Schematic view of a (100) surface. The red circles are depicting occupied adsorption sites (bridge sites) Arrows indicate the adsorbate–adsorbate interaction between the adsorbed particles with the assigned name. Some examples for pair interactions and some higher order many-body interactions are given.

A systematically different approach starts at the microscopic level. Here a given set of ordered structures is calculated with a higher level of theory. Usually this is done at the level of electronic structure calculations. The resulting total energy of these calculations is then expanded in terms of a truncated CE. These higher level calculations capture intrinsically the underlying interactions included as interaction parameters in the CE. Afterwards a fitting process is performed and the interaction energy contributions are used as fit parameters to express the total energy of each structure by means of the CE. The parameters are determined such that they minimize the average fitting error, to express the total energy of each configuration as accurate as possible in comparison to the energy provided by the electronic structure calculations.[90] This procedure is what is usually summed under the name *first-principles cluster expansion (FP-CE)*. Two crucial aspects determine the reliability of this approach: First, a set of fitting

---

[^1]: Please note that for now contributions by vibrations are neglected and thus we are not approximating the free energy of the system, but only the configurational energy. For a brief discussion on that see subsection 5.4.1)

parameters, ergo interaction parameters, has to be chosen. Secondly, how many input structures for the fit from the higher level theory are necessary to ensure a certain accuracy in the total energies? The next sections will address these questions.

5.2 Finding the Important Interaction Parameters − Cross Validation

The identification of the set of interactions to take into account as fitting parameters is usually driven by a combination of physical consideration and a quite systematic expansion. This means, that one usually assumes that pair interactions of particles in close proximity will represent rather high energy contributions and that surely some trio interactions will be important for the stabilization of certain ordered overlayer structures as well. However, which higher order many-body interactions are important or not and up to which distance, for example, trios should be considered is a difficult task to decide. A quite systematic way to select the important interactions is presented by the approach of Cross-Validation (CV)[91].

This term usually denotes the so-called *Leave-One-Out Cross-Validation (LOO-CV)*, but also other approaches like *Monte-Carlo-Cross-Validation (MC-CV)* are frequently used[92], which overcome the problem of the *asymptotic inconsistency* of the LOO-CV.[93] This term describes the fact that when the size of the fit database goes to infinity the probability to pick the exact CE must equal unity − which is not the case for LOO-CV.

The underlying principle of the cross validation approach is to define an estimator, the *CV-score*, for the *predictive power* of the used expansion with a given set of interaction terms. This estimator is determined by performing the fitting process to determine the best set of values for the set of interaction parameters only for some configurations (the training set) of the higher level of theory dataset. The resulting values for interaction parameters are then used to calculate a predicted configurational energy for all remaining configurations (the validation set), which have not been used for the fitting procedure. By this, one can get the error in prediction as energy difference of the CE-predicted total configurational energy and the known total energy from the higher-level calculation. In the end, an average of all errors is calculated and after normalization taken as the CV-score for this set of interaction parameters.

As already mentioned, in its simplest form, the LOO-CV, this means using all but one high level calculations for the fit and validating against the configuration that has been left out. This is repeated until each structure have been excluded from the fitting once and all errors in the total configurational energy are added up, averaged and normalized. The LOO-CV Score is thus defined as:

$$\text{LOO-CV} = \sqrt{\frac{1}{N} \sum_{i=1}^{N} [E_{\text{conf.}}^{\text{CE}}(i) - E_{\text{conf.}}^{\text{DFT}}(i)]^2} \quad , \tag{5.2}$$

in which N represents the total number of different ordered configurations calculated with the higher theory method, $E_{\text{conf.}}^{\text{DFT}}(i)$ is the calculated energy by the higher method (here:

DFT), and $E_{\text{conf.}}^{\text{CE}}(i)$ is the predicted total configuration energy by the fitted cluster expansion (CE), respectively. For each structure i the fit is performed using the remaining $(N-1)$ configurations and a least-squares fitting routine of the resulting expansion matrix to calculate the interaction parameters.

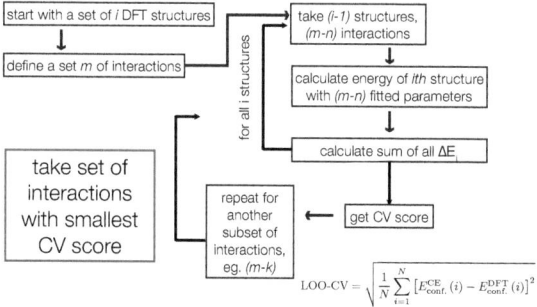

Figure 5.2: Flowchart for the iterative procedure used to determine the optimal set of lateral interaction parameters to be included into the Cluster Expansion by determining the Cross Validation Score of each single permutation.

A general flowchart of the procedure employed in this work to determine the optimal set of lateral interactions is shown in Fig. 5.2. Initially one starts with a set of m interaction terms and selects a subset of $(m-n)$ interactions. For this the CV Score is calculated by the procedure described above according to Eq. (5.2). This is repeated for all possible subset permutations with a subset size $3^{\text{II}} \leq k \leq m$ for the interactions given in Fig. 8.7. The one subset that minimizes the CV Score is taken for all subsequent calculations.

5.3 Is the Fit Base Complete? – Ground-State Validation

The quality of a cluster expansion is not only determined by a minimized prediction error but also by its ability to predict the correct ordered ground state structures[95]. For this reason, since structures found in experiments at finite temperatures are obviously (meta) stable structures, these configurations should be included into the fit. In addition, one would like to include the set of most stable structures as predicted by the higher order theory into the fit. Although usually calculated within the higher level theory for $T = 0$ K, one would hope the real system at finite temperatures to be a statistical mixture of the ground state structures.

[II] The selection of the minimum number of interactions to be taken into account is motivated by acknowledging the observation that for minimal ordering behavior with clustering at least three parameters are needed on (100) surfaces[94].

5.3.1 The Formation Energy

A convenient measure for of stability of adsorbate structures is the *formation energy*, $\Delta E_{\text{form.}}$. For adsorbate systems, the formation energy decribes the excess energy with respect to the pure constituents (in the example system of chapter 8 these are CO in gas phase and the clean Pd(100) surface). The following part will explain the method of the ground state validation procedure employed in this thesis for this example system.

The formation energy for an adsorption configuration of CO on a Pd(100) surface is given by:

$$\Delta E_{\text{formation}} = \frac{1}{N_{\text{adsorption sites}}} \left[E^{\text{total}}_{\text{CO@Pd(100)}} - \theta \cdot E^{\text{total}}_{(1\times 1)\text{CO@Pd(100)}} - (1-\theta) \cdot E^{\text{total}}_{\text{Pd(100)}} \right] \quad (5.3)$$

In this equation, the total energy of the specific adsorbate configuration is given by $E^{\text{total}}_{\text{CO@Pd(100)}}$, $N_{\text{ads. sites}}$ gives the number of adsorption sites in the surface unit-cell and the coverage θ for a configuration with N_{CO} atoms adsorbed is defined as $\theta = N_{\text{CO}}/N_{\text{ads. sites}}$.

The energy of a clean surface with no adsorbates is given by the value of $E^{\text{total}}_{\text{Pd(100)}}$. The last term, $E^{\text{total}}_{(1\times 1)\text{CO@Pd(100)}}$ describes the energy of a configuration with monolayer coverage, i.e. in principle that would be the situation where all available adsorption sites of one kind in the surface unit-cell are occupied. However, for the particular system described here, the coverage is normalized to the number of available hollow sites in the $p(1 \times 1)$ surface unit cell, which introduces a factor of 1/2 to the total number of available adsorption sites. For brevity, however, this factor is neglected in the following explanations.

The definition in Eq. (5.3) can be related to the binding energy of a configuration as given by a total energy calculation simply by

$$\Delta E_{\text{formation}} = \theta \cdot \left[E^{\text{bind}}_{\text{CO@Pd(100)}} - E^{\text{bind}}_{(1\times 1)\text{CO@Pd(100)}} \right] \quad . \quad (5.4)$$

Defined in this way, $\Delta E_{\text{form.}}$ describes the relative stability of any configuration with respect to the phase separation into a fraction θ of the full monolayer configuration and a fraction $(1-\theta)$ of the clean surface.

As shown in figure, Fig. 5.3, one can now plot the formation energy versus coverage for several ordered structures. In this way, the so-called *DFT ground state line* or *convex hull*[82] is defined. In the example shown here,

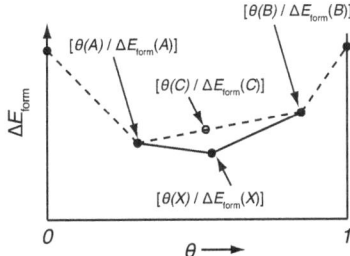

Figure 5.3: Convex hull ground state line with interpolated value (explanation see text).

initially there may be the configuration A with $\theta(A)$ and corresponding formation energy $\Delta E_{\text{form.}}(A)$ and another configuration B with $\theta(B)$ and $\Delta E_{\text{form.}}(B)$, respectively. This situation is presented by the dashed line in Fig. 5.3. Every configuration that constitutes a mixture between these two ground state configurations A & B will have a formation energy that lies on the dashed line between them[86], namely the one easily derived by linear interpolation:

$$\Delta E_{\text{form.}}(C) = \frac{\theta(C) - \theta(A)}{\theta(B) - \theta(A)} \cdot \left[\Delta E_{\text{form.}}(B) - \Delta E_{\text{form.}}(A) \right] + \Delta E_{\text{form.}}(A) \qquad (5.5)$$

Checking the formation energy of all possible structures enables one then to check, whether the CE would predict configurations with energies that fall below the thus defined *convex hull* given by the dashed line. This situation is schematically represented by the configuration X in figure Fig. 5.3. Obviously configuration X represents an important ordering motif as its formation energy is lower than the convex hull. The way to deal with a situation like this is of course, within an iterative approach for the validation of the CE, to incorporate this very motif into the ground state training set and iteratively repeat the procedure of fitting, cross-validation and ground state validation. This way, step by step a new ground state line (solid line) is iteratively defined. This means eventually that one will include all motifs that are low lying in energy ideally also as a DFT value into the first principles database. Finally one would converge to a situation, in which all possible structures calculated by the CE would lie higher in formation energy and are thus compatible with the stability know at the DFT level.

5.3.2 Ground-State Validation and Direct Enumeration

The formally correct way to check for this for a problem of adsorbates on a surface is simply to generate all possible configurations of adsorbates within all possible unit cells for $0 \leq \theta \leq 1$ by a permutation algorithm. This approach is the so-called *direct enumeration*. However, to enumerate all sorts of unit cell sizes and shapes is of course not trivial and also yet computationally infeasible. Therefore, for the purpose of the work presented here only two different unit cell sizes in different shapes were taken and all possible adsorbate configurations for $0 \leq \theta \leq 1$ were generated by a combinatorial algorithm. It was critical here of course to which reference structure the surface coverage refers to for a monolayer (1 ML) coverage and for this value the $p(1 \times 1)$ structure was taken as reference, as already mentioned.

Finally, the ground state validation was not carried out in an iterative way to improve step by step the cluster expansion, but more as a measure of predicting power to get an idea for which coverages the found expansion breaks down. It is well known that not always the cluster expansion with the best CV score is able to predict the correct ground state structures. In these case, one approach can be to include different configurations with different statistical weights[96]. This would be a step for further improvement of the cluster expansion presented here.

5.4 Sources of Uncertainties in the Predictions – Some Remarks about the Accuracy

5.4.1 Uncertainties in the first-principles energies data

First and most important for the energetics resulting from a cluster expansion is the fact that the first-principles energetic calculations are not without approximations.

Vibrational Contributions to the Free Energy

Formally, Eq. 5.1 should not just take into account the static total energy arising by a certain adsorbate configuration, but be written in terms of the binding free energy of the system, $F_{bind.}(T)$, which contains also contributions from vibrations. The average binding free energy can be split into the static binding energy from Eq. 5.1, $E_{conf.}^{tot.}$, and the vibrational contributions $F_{bind.}^{vib.}(T)$:

$$F_{bind.}(T) = E_{conf.}^{tot.} + F_{bind.}^{vib.}(T) \quad . \tag{5.6}$$

In the following, a short methodological approach to this contribution will be sketched [97], in which this energy contribution by the vibrations is derived from the phonon density of states, $\sigma(\omega)$. Using the phonon density of states the vibrational free energy can be written as

$$F_{bind.}^{vib.}(T) = \int F(T,\omega)\sigma(\omega)\,d\omega \quad , \tag{5.7}$$

where

$$F(T,\omega) = \hbar\omega\left(\frac{1}{2} + \frac{1}{\exp[\beta\hbar\omega]-1}\right) - k_B T \left[\frac{\beta\hbar\omega}{\exp[\beta\hbar\omega]-1} - \ln\left(1-\exp[-\beta\hbar\omega]\right)\right] \tag{5.8}$$

is the vibrational free energy of an harmonic oscillator of frequency ω, k_B is the Boltzmann constant and $\beta = 1/(k_B T)$. Similar to the evaluation of the pure binding energy by differences of total energy calculations, see Eq. 2.8, the vibrational binding free energy contribution can be written as difference between total vibrational free energies of the complete system and its sub-systems. The difference is then taken between the surface phonon density of states for the adsorbate covered and clean surface, as well as the vibrational frequencies of the gas phase reference state for the adsorbate.

However, in all further calculations employing the CE approach, only static contributions are taken into account and the vibrational contributions will be neglected. This introduces some error to the energies evaluated for different configurations, of course, and therefore represents an uncertainty. In a rough estimate, based on former work [18], it seems safe to state that these contributions will be rather small. They have been estimated to be of the order of ≤ 5 meV for the lateral interaction energy and of the order of ≤ 30 meV for the on-site binding energy contributions for the similar system Pd(100) and O_2. The dominant part causing the rather high vibrational contribution to the on-site binding was identified with the shift in the vibrational frequency between the stretch vibration of the free O_2 molecule and the changed stretch vibration mode of the adsorbate–substrate bond. Considering that the gas phase molecule of the current study, CO, adsorbs molecularly and not dissociatively, one may argue that this effect will be smaller than for the situation of the dissociative adsorption of oxygen.

Uncertainties by the approximate xc functional

As stated already in section 3.2, the modern density-functional theory absorbs all unknown energy contributions into the xc-functional. Since the exact mathematical form of this functional is unknown, it is approximated in all practical calculations.

Through this approximate xc-functional, naturally, a fundamental source of uncertainty for the predicted total energies is introduced. A common argumentation proceeds along the line that normally only energy-*differences* of total energies are given a physical interpretation and therefore uncertainties in the absolute total energy values would at least partly cancel out each other. This is commonly referred to as cancellation of errors. In the comparison to experimental values for geometries and energetics it is evident that the GGA-PBE[60,61] xc-functional employed throughout the calculations presented here performs quite well for a wide range of systems (ranging from molecules to solids) while having a direct connection to physical principles[98].

However, it must be stated that the uncertainties introduced by the xc-functional cannot be as systematically explored within this work as this is possible for the other parameters of the DFT calculations.

Uncertainties by the numerical procedures

The influence of several parameters that need to be selected for a DFT calculation as implemented in the CASTEP plane wave code on the respective quantity of interest has been carefully tested in the section on convergence tests, see Appendix A.

These parameters are in particular:

- geometrical parameters, such as slab thickness – the number of layers to model the surface and bulk part of the metal, and the vacuum separation between periodic images,
- numerical parameters, such as the convergence criteria for the different self-consistent iterative procedures,
- sampling parameters, such as the number and distribution of k-points for the sampling of the Brilluoin zone,
- cut off energy of the plane wave expansion,
- quality of the employed pseudopotential descriptions.

5.4.2 Uncertainties in the cluster expansion procedure

Uncertainties in the cluster expansion and the extracted interaction parameters arise naturally from the truncation of the cluster expansion. In addition, the limitation in the number

5.4. Sources of Uncertainty – Accuracy Considerations

of input configurations used in the fitting process introduces another source of uncertainty.

5.4.3 Concluding Remarks

As extensively discussed in a recent paper by Zhang et al.[99], the major effect of all uncertainties in the DFT procedure mainly affect the accuracy of the on-site energy value contribution to the total energy. The main reason for that is the unsufficient description of the gas-phase reference state of the adsorbates by electronic structure codes for bulk/surface systems.

In addition, one has to keep in mind that the method of cluster expansion CE is just a mathematical fitting procedure. The final choice whether the predicted parameters do have a physical meaning or are just occasionally caused by some mathematical instability in the fitting procedure lies in the hand of the user. For the interpretation one needs to take into account the physics of the problem and knowledge about the system under investigation from other methods.

Direct Representation of Lateral Interaction Parameters

However, the beauty of this method lies in the fact that it provides direct energy values for the different kinds of considered interaction parameters. This is, of course, also the largest weakness of this approach. At first, only lateral interactions already considered in the basic pool of interaction parameters which is providing the starting point for the CV score evaluation will be taken into account. The method itself is in the current implementation not able to find important parameters that have not been defined beforehand for itself. Therefore, the set up of the initial pool of parameters requires some physical intuition. Secondly, by the strict definition via distances to neighboring atoms which are adsorbed site specific, the method is not able to cover relaxation effects which would lead to geometry distortion.

Chapter 6
Methods - Embedded-Atom-Method (EAM)

ONE can derive semi-empirical interatomic potentials from several starting points. One approach can be to start with a set of values from some measurable material property. These values are then connected to some internal variables by a theory. The internal parameters of this theory are finally adapted in a fitting process such that the experimentally known values are optimally reproduced. Naturally one needs a starting point for this procedure and it is by no means unique what this starting point should be. Therefore, in the following description of the semi-empirical interatomic potential that has been used through this work, a focus will be laid on presenting the functional form and its fitting parameters.

As the special focus of this thesis is placed on the description of lateral interactions between adsorbates some remarks about their inclusion in the semi-empirical potential method presented in the following, namely the so-called *embedded-atom-method* (EAM)[100–106] will be made.

We will see that by means of the usage of the electron density as a material specific property the potential formulation includes many-body effects implicitly to a certain extent. In this respect, the semi-empirical description and the description by the higher order electronic structure method are somewhat similar. In addition, the fitting procedure employed for the parameterization ensures that effects of lateral interactions are included in the potential as well, simply by the fact that lateral interactions will affect certainly some of the material properties used in the fit set employed for the potential parameterization.

6.1 Total Energy in the Embedded-Atom-Method

The example system for the application of this semi-empirical potential is composed of copper islands on a silver (100) surface. The following more detailed description of the construction of the employed functional is therefore oriented on the paper by William *et al.* in which the EAM parametrization for this system has been presented originally.[107]

In the EAM formalism the total energy of a system is expressed in the form:

$$E_{\text{tot}}^{\text{EAM}} = \frac{1}{2}\sum_{ij} V_{ij}(r_{ij}) + \sum_{i} F_i(\bar{\rho}_i) \quad , \tag{6.1}$$

in which $V_{ij}(r_{ij})$ represents the pair interaction energy between atoms i and j separated by distance r_{ij} and F_i is the embedding energy of atom i as a function of the host electron density $\bar{\rho}_i$. This latter quantity is given by:

$$\bar{\rho}_i = \sum_{j \neq i} \rho_j(r_{ij}) \quad , \tag{6.2}$$

in which $\rho_j(r)$ is the electron density function assigned to atom j. Therefore, an EAM description of a system that consists only of one element, A, requires three functions, namely,

- the pair potential, $V_A(r)$, which represents mainly the electrostatic particle interactions,
- the density function assigned to each atom A at position r, $\rho_A(r)$, that describes the electron density distribution,
- and finally the embedding function, $F_A(\bar{\rho})$, which is a funtion of the host electron density. The host density in turn is generated by a superposition of the atomic density functions described before, compare equation Eq. 6.2.

Due to the fact that the host electron density is represented within this formalism as a superposition of atomic densities, naturally this description is best suited for systems with an almost homogenous electron density distribution. Therefore, EAM potentials are in particular suited for the atomistic description of metallic systems.

For the parametrization of a binary system containing two different species, one of course needs three functions for each species. The cross-interaction between the particles of different kinds is then usually covered in an additional special pair potential, which in total gives seven functions that needs to be parameterized. The host electron density is generated in the binary case as a superposition of two different types of atomic density functions, but usually not parameterized separately.

6.1.1 Representation of Lateral Interactions

The interactions between particles, either within the bulk or as adsorbates on a surface, are covered in two ways. Obviously, pair interactions are accounted for directly via the pair interaction function. Usually, this particular potential is subject to a cut-off function, which cancels its effects above a certain cut off radius (for a closer description see equation Eq. 6.7 and its explanation below).

All many-body interactions are intrinsically covered by the embedding formalism employed in this interatomic potential method. Since these semi-empirical potentials are usually derived by a fitting process to certain experimental and/or first-principles data, it is ensured that these fit set parameters, which in turn are intrinsically covering many-body effects, are reproduced by tuning the embedding function. So, by construction, the effects of

6.1. Total Energy in the Embedded-Atom-Method

many-body lateral interactions between the particles of the system are included indirectly. For a binary system, respectively, the fitting process of the embedding function and the fact that it depends on the superposition of the atomic electronic densities of *all* particles again covers higher order interactions also indirectly.

6.1.2 Uniqueness of the EAM Energy Function

The partitioning of E_{tot} into pair interactions and the embedding energy is not unique, see Ref. [105] and references therein. For the mathematical construction of an EAM potential this statement represents an important point. $E_{\text{tot}}^{\text{EAM}}$ as defined above in equation Eq. 6.1 is invariant under certain transformations, namely:

$$F_i(\bar{\rho}) \;\mapsto\; F_i(\bar{\rho}) + g_i \bar{\rho} \quad, \tag{6.3}$$

$$V_{ij}(r) \;\mapsto\; V_{ij}(r) - g_i \bar{\rho}_i(r) - g_j \bar{\rho}_r(r) \quad, \tag{6.4}$$

where i,j are the atomic indices and the g_x are arbitrary constants. Also the density functions can be scaled by an arbitrary factor, k as long as the argument of the embedding function is scaled with the same factor appropriately, that is:

$$\rho_i(r) \;\mapsto\; k \rho_i(r) \tag{6.5}$$

$$F_i(\bar{\rho}) \;\mapsto\; F_i(\bar{\rho}/k) \tag{6.6}$$

Given these ambiguities, it is important to carefully define the functional form and fitting parameters used for an embedded-atom potential to avoid confusion. This will be done in the remaining part of this chapter for the Ag–Cu alloy system (which will be the example system in chapter 9).

6.2 The Parametrization of the EAM Potential for pure Silver

As a first step in the parametrization of an EAM potential, one needs to decide for a functional form for the individual elements of it.

The pair potential

In the potential by Williams et al.[107] used in this study, the pair potential function $V(r)$ is represented by a superposition of two Morse functions, $\mathcal{M}$:

$$V(r) = V(r)_M \equiv \left[E_1\mathcal{M}\left(r,r_0^{(1)},\alpha_1\right) + E_2\mathcal{M}\left(r,r_0^{(2)},\alpha_2\right) + \delta\right]\psi\left(\frac{r-r_c}{h}\right) \quad , \tag{6.7}$$

if $r \geq r_s^{(1)}$ and

$$V(r) = V_M\left(r_s^{(1)}\right) + V_M'\left(r_s^{(1)}\right)\left(r - r_s^{(1)}\right) + \frac{1}{2}V_M''\left(r_s^{(1)}\right)\left(r - r_s^{(1)}\right)^2$$

$$+ \frac{1}{6}V_M'''\left(r_s^{(1)}\right)\left(r - r_s^{(1)}\right)^3 + \sum_{n=1}^{5} S_n H\left(r_s^{(n)} - r\right)\left(r_s^{(n)} - r\right)^4 \tag{6.8}$$

if $r < r_s^{(1)}$.

In the expression for the pair potential given above, the following notations have been used:

- $\mathcal{M}(r,r_0,\alpha) = \exp[-2\alpha(r-r_0)] - 2\exp[-\alpha(r-r_0)]$, is the used Morse function;
- $H(x)$ is a unit step function and $\psi(x)$ is a cut off function, which is equal 0 if $x \geq 0$ and $\psi(x) = x^4/(1+x^4)$ otherwise;
- the set $E_1, E_2, r_0(1), r_0(2), \alpha_1, \alpha_2, \delta, r_c, h$ and $\left\{r_s^{(n)}, S_n\right\}_{n=1,\ldots,5}$ are the fitting parameters.

For the description of the electron density distribution the following functional ansatz is parametrized:

For $r \geq r_d^{(1)}$, it is:

$$\rho(r) = \rho_d(r) \equiv \left[A\exp\left[-\beta_1\left(r - r_0^{(3)}\right)^2\right] + \exp\left[-\beta_2\left(r - r_0^{(4)}\right)\right]\right]\psi\left(\frac{r-r_c}{h}\right) \tag{6.9}$$

and for $r < r_d^{(1)}$ one sets:

$$\rho(r) = \rho_d\left(r_d^{(1)}\right) + \rho_d'\left(r_d^{(1)}\right)\left(r - r_d^{(1)}\right) + \frac{1}{2}\rho_d''\left(r_d^{(1)}\right)\left(r - r_d^{(1)}\right)^2 +$$

$$\frac{1}{6}\rho_d'''\left(r_d^{(1)}\right)\left(r - r_d^{(1)}\right)^3 + Q_1\frac{\left(r - r_d^{(1)}\right)^4}{1 + 9\left(r - r_d^{(1)}\right)^2} + \tag{6.10}$$

$$Q_2 H\left(r_d^{(2)} - r\right)\left(r - r_d^{(2)}\right)^4 \quad .$$

The set of fitting parameters for the density expression is: A, $r_0^{(3)}$, $r_0^{(4)}$, β_1, β_2, $r_d^{(1)}$, $r_d^{(2)}$, Q_1, Q_2.

6.2. The Parametrization of the EAM Potential for pure Silver

The cut-off function

$\psi(x)$ ensures that both position dependent functions as well as their derivatives up to the second order turn smoothly to zero at the cutoff distance r_c, which imposes an important property for the setup of simulation ensembles. Knowing the cut-off distance of the underlying potential enables one to include as many atoms into the simulation cell as necessary to describe all interactions around a certain interesting point in space of the simulation cell, but also allows to keep the simulation cell as small as possible. This is due to the fact, that atoms which are separated in space more than the cut off distance of the potential obviously do not interact with each other anymore.

The embedding function

To express $F_\rho(\bar{\rho})$ one starts with a polynomial ansatz:

$$F_\rho(\bar{\rho}) = F^{(0)} + \frac{1}{2}F^{(2)}(\bar{\rho}-1)^2 + \sum_{n=1}^{4} q_n (\bar{\rho}-1)^{n+2} \qquad (6.11)$$

For the case $\bar{\rho} < \bar{\rho}_1$ one directly takes the equation above, Eq. 6.11, whereas for the other limiting case, namely $\bar{\rho} > \bar{\rho}_1 > 1$ one expresses the embedding function as:

$$F(\bar{\rho}) = F_\rho(\bar{\rho}_1) + F'_\rho(\bar{\rho}_1)(\bar{\rho}-\bar{\rho}_2) + \frac{1}{2}F''_\rho(\bar{\rho}_1)(\bar{\rho}-\bar{\rho}_2)^2 + \frac{1}{6}F'''_\rho(\bar{\rho}_1)(\bar{\rho}-\bar{\rho}_2)^3 \qquad (6.12)$$

6.2.1 The Parameterization Procedure Used

In the parameterization approach employed by Williams et al. [107] Two of the fitting parameters, namely A (see equation Eq. 6.9) and E_1 (see equation Eq. 6.7), are eliminated by normalizing the density function $\rho(r)$ to $\bar{\rho} = 1$ at the lattice parameter $a = a_0$. This reduces the number of parameters by two. A further reduction in the set of fitting parameters can be achieved by expressing the coefficients $F^{(0)}$ and $F^{(2)}$ (in equation Eq. 6.11) in terms of the experimental cohesive energy E_0 and the bulk modulus B. Furthermore, q_1 is determined using the boundary condition $F_\rho(0) = 0$. Thereby, that leaves only ρ_1 next to q_2, q_3, q_4 as fitting parameters for $F(\bar{\rho})$.

For the parameters $\bar{\rho}_1$, $r_s^{(s)}$ and S_n arbitrary parameters have been selected to influence the repulsive part of the binding curve in the desired way. Similar, for $r_d^{(n)}$ and Q_n parameters have been chosen such that negative values of $\rho(r)$ at short atomic separations are prevented. By this reduction, in total 30 adjustable parameters are present within the final functional form. However, in practice for the parametrization for the Ag part described here, only 15 parameters are fitted to the experimental properties of the pure phase.[107] The procedure for the parameter selection presented here ensures automatically an exact fit of the potential to a_0, E_0 and B.

As an experimental database for the fit besides the already included a_0, E_0 and B of Ag, the elastic constants C_{ij}, the relaxed vacancy formation E_v^f and migration E_v^m energies, the phonon frequencies at the zone-boundary point X and the intrinsic stacking fault energy γ_{SF}

are used. Also included was the experimental thermal expansion factor of fcc Ag at 1000 K to a small weight.

The first-principles database contained per atom energies as a function of volume generated with the VASP code[108] for several bulk structures of Ag. It is important to note here that the employed exchange-correlation functional was the local density approximation (LDA) in contrast to the DFT calculations that were performed within our study which used the generalized-gradient approximation (GGA) functional as published by Perdew et al. [60,61]. Another important aspect, which needs to be kept in mind for all comparisons between first-principles interatomic distances and EAM interatomic distances, is presented by the fact that the first-principles volume dependent energy functions need to be scaled to account for the mismatch in the equilibrium lattice parameters of any DFT xc-functional and the experimental lattice constant used in the EAM parametrization before.

The parameterization for the Cu part of the binary alloy potential was taken from a previously published EAM potential parameter set by Mishin et al. given in Ref. [109]

6.3 The Cross-Potential for the Ag–Cu Interaction

As already stated, the cross-interaction for a binary alloy EAM potential is given within this formalism as a pair potential $V_{AB}(r)$. The functional form employed is again a Morse function identical to the one presented already for the pure compound.

$$V_{AB}(r) = E_1 \Big[\exp(-\alpha\beta(r-r_0)) \\ -\alpha \exp(-\beta(r-r_0)) + \delta \Big] \quad (6.13) \\ \cdot \psi\left(\frac{r-r_c}{h}\right)$$

As the specific system for which this EAM potential is intended shows a wide miscibility gap in the solid phase[110,111], the cross-interaction fitting database was generated from first-principles with the VASP code and the LDA exchange correlation functional. It contained seven artifical intermetallic ordered phases of silver and copper for which energy-volume relations have been extracted. After the necessary scaling to account for the different equilibrium lattice constants of experiment and LDA DFT calculations, the fitting was performed such that the mean-squared deviation between the EAM predicted and the first-principles predicted formation energies for all atomic

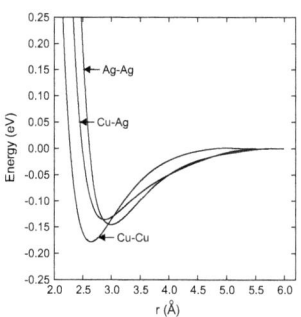

Figure 6.1: Shown in this plot are the three resulting pair interaction functions resulting from the EAM parametrization for the Ag-Cu alloy EAM potential that was used in this thesis. Picture taken from Ref. [107]

volumes was minimized. Fitting parameters were the set of: E_1, r_o, α, β, δ, r_c and h. In addition, the scaling factors of the electronic density functions of Ag, s_{Ag} and the transformation coefficients g_{Cu}, g_{Ag}[105] were also used as fitting parameters.

6.4 On Benefits and the Validation of the EAM Potential

EAM potentials come with the great benefit, that once thoroughly parametrized for a special system, their computational evaluation is very cheap and lies in the order of a simple pair potential.[105] This allows the treatment of very large systems with millions of atoms with a much smaller computational effort compared to electronic structure calculations.

The validation of an EAM potential is in principle possible by comparison between predicted energy values calculated within the EAM formalism for a set of structures not used for the EAM parametrization and the energy values of the same structures calculated by a higher order theoretical method, such as density-functional theory. Careful attention must be paid to the underlying parametrization database as for example for different equilibrium lattice parameters. To compare geometries, interatomic distances need therefore to be scaled, for instance. Another important aspect is the employed exchange-correlation functional, in case the EAM parametrization contains also datapoints from first-principles calculations. If compared to DFT calculations that are making use of a different exchange-correlation functional, this can lead to a constant offset between the datapoints. Naturally, the employed exchange-correlation functional also affects the predicted theoretical equilibrium lattice constant which appropriately needs to be taken into account in the scaling procedure.

The EAM potential for this thesis has been taken in tabulated form from the Interatomic Potential Database at NIST [112].

Chapter 7
Methods - Optimization Techniques

THE task that naturally comes before any meaningful parameterization or extraction of lateral interaction parameters is to find the minimum energy configuration within the configurational space of the given system. Obviously this includes first an energy minimization of the given starting structure with respect to the atomic coordinates $\{\mathbf{R}_{a,i}\}$. It is important to keep in mind, that this structure minimization usually can converge only to the next *local* minimum configuration on the potential energy surface (PES) and not to the absolute gobal minimum configuration of the given system at zero temperature.

However, also local minima configurations can be of interest, because depending on the energy barrier between the global and most stable minimum configuration and a certain local minimum the latter one can be accessible and populated at finite temperatures according to the rules of statistical thermodynamics.

In order to find the optimum configuration two things are needed. At first, a method that yields the total energy of any configuration of atomic coordinates. The methods employed for the total energy evaluation for the systems investigated here have been presented already in the chapters before. Secondly, a method is needed that generates a search direction for the generation of a new configuration within an iterative approach based on the information given by the local PES environment of the current configuration. The current structure is then updated accordingly until convergence in an iterative approach is reached. Some methods used in this work to perform this task will be described as part of this chapter.

Finally, a method to extract information about the barrier heights between local minima will be presented. This information is useful for ways to explore the PES. The methods to explore the PES of this thesis can be roughly categorized into two fields, on the one hand the PES will be explored randomly, on the other hand the system's time evolution will be sampled and the PES will be explored according to Newton's laws while making use of information from transition state theory to accelerate the exploration.

7.1 Structure Optimization by Energy Minimization

For all energy minimization techniques presented here, the update of the configurations is performed by taking into account the curvature of the PES. For each atom position, the first derivative of the total energy with respect to the coordinates is needed, i.e. the *forces* acting on each atom

$$\mathbf{F}_\alpha(\mathbf{R}_\alpha) = \frac{\partial E}{\partial \mathbf{R}_\alpha} \quad . \tag{7.1}$$

7.1.1 Steepest Descent and Conjugate Gradient

In the most simple local minimization method, the *steepest descent* method, one iteratively follows the curvature of the potential energy surface strictly downhill. In each iteration step the atomic coordinates are updated according to the forces acting on them. This is

$$\mathbf{R}_{\alpha,i+1} = \mathbf{R}_{\alpha,i} + \gamma_i \mathbf{F}_\alpha \left(\{\mathbf{R}_{\alpha,i}\} \right) \quad , \tag{7.2}$$

where γ_i is a step width parameter (not necessarily constant over all iteration steps) to adjust the step width. Selecting a too small value will result in very slow convergence, however, a γ_i that is too large leads to oscillations around the minimum in this method.

For the *conjugate gradient* minimization a search direction $\mathbf{G}_{\alpha,i}$ for the minimization problem is constructed according to the following equation[113], using also the forces acting on the atoms as information about the search direction

$$\gamma_i = \arg\min_\gamma E \left[\mathbf{R}_{\alpha,i} + \gamma \mathbf{G}_{\alpha,i} \left(\{\mathbf{R}_{\alpha,i}\} \right) \right] \quad . \tag{7.3}$$

Here, the expression $\arg\min_\gamma E$ is describing the argument γ which minimizes the energy functional. This argument value is then used in the next iteration to update the atomic coordinates according to

$$\mathbf{R}_{\alpha,i+1} = \mathbf{R}_{\alpha,i} + \gamma_i \mathbf{G}_{\alpha,i} \quad . \tag{7.4}$$

In contrary to the steepest descent method, also information about previous steps are included in the update of the atomic coordinates. In this way, a faster and more robust convergence can be achieved that follows the PES downhill and is always approximately perpendicular — «conjugate» — to the previous search directions. The information of previous steps is taken into account by adding a fraction of the previous search directions to the atomic forces of the actual position according to

$$\mathbf{G}_{\alpha,i} = \mathbf{F}_{\alpha,i} + \beta_i \mathbf{G}_{\alpha,i-1} \quad . \tag{7.5}$$

There exist different ways to construct the corresponding technical parameter β_i which determines the mixing fraction. The construction of the search directions $\mathbf{G}_i$ is based on the assumption that the PES is harmonic. However, this also means that in case the system is far away from the harmonic region, the search directions constructed by the conjugate gradient scheme are somewhat unreasonable.

Either one of these two schemes has been used for the structure minimizations of copper island structures performed in chapter 9 of this thesis.

7.2. Structure Optimization by Global Optimization

7.1.2 The Broyden-Fletcher-Goldfarb-Shanno (BFGS) Method

In all geometry optimizations on the level of electronic structure theory a so-called *quasi-Newton* scheme[114] that takes additional information of the second derivative of the PES into account was used to converge to the local minimum structure. The BFGS method constructs the new search directions by employing the Hessian matrix $H_{\alpha\beta}$ of the system. The new $\mathbf{G}_{\beta,i}$ are generated by solving the Newton equation

$$\sum_\beta H_{\alpha\beta} \mathbf{G}_{\beta,i} = \mathbf{F}_\alpha \left(\{\mathbf{R}_{\alpha,i}\}\right) \qquad . \tag{7.6}$$

For a perfectly harmonic PES and an exactly known Hessian, the local minimum would be found within one line search. In practice, obtaining the exact Hessian of a system can be a quite cumbersome task, so that usually this matrix is successively approximated within each iteration step. However, since more information of the PES is taken into account for the minimization the BFGS method is usually the most effective one of the three presented schemes for local minimization so far. As for the conjugate gradient method, also for the BFGS scheme being in an area of the PES where the harmonic approximation holds is crucial for a satisfying performance.

The BFGS algorithm was the method exclusively used for all geometry optimizations performed within the framework of density-functional theory using the CASTEP[74] computer code.

7.2 Structure Optimization by Global Optimization

In order to find the global minimum configuration for a given system, one has to use a technique that samples the phase space of the PES and is not constrained to a minimization. The problem with the usage of energy minimization procedures lies, as already stated, in the fact that only relaxations into the next local minimum structure are possible. This is caused by the fact that the local information — or information of a close proximity region — for a given configuration uniquely guides the system to this next local minimum. This fact can be contracted to the statement, that the starting configuration defines the accessible configurations.

In both global optimization techniques presented in the following, this local information is replaced by a random element.

7.2.1 Simulated Annealing

Simulated annealing[115] represents a global optimization scheme that uses a principle from statistical mechanics and metallurgy for a multidimensional optimization scheme. A necessary condition is the existence of a «cost» function which calculates a value to be minimized during the optimization process. In global optimization problems of molecular structures, this function is normally taken to be the total energy, E^{total}, of the particular atomic configuration.

A typical simulated annealing run starts with an arbitrary configuration of total energy E_1 and a given assumed starting temperature, T_{start}, of the system. By randomly displacing one or more atoms, a new structure is generated and the total energy, E_2 for the new structure is calculated and compared to the one of the starting configuration. If the energy of the configuration has decreased, i.e. $\Delta E = E_2 - E_1 < 0$, the new configuration is accepted and represents the new starting configuration for the next iteration. In the opposite case that the energy has increased, $\Delta E > 0$, the new configuration is accepted with a probability $P(\Delta E)$ according to a Metropolis criterion[116], i.e. $P(\Delta E) = \exp(-\Delta E/k_B T)$, where T is equal to T_{start} in the beginning. According to Metropolis this generates a canonical ensemble of atomic configurations for the given temperature, i.e. the statistically most likely configuration with the highest probability at a given temperature.

Consequently, at zero temperature only new structures that would lower the energy and therefore approach the minimum configuration would be selected. However, this temperature setting bears the risk to get stuck in the wrong local minimum and not to find the global minimum configuration. A finite temperature setting, on the contrary, allows for certain uphill steps and thus enables the system to find its way out of local minima, thereby exploring large areas of the configurational phase space during the sampling. In the simulated annealing procedure, the system is usually sampled starting from a high starting temperature and then successiveley cooled down according to an annealing schedule.

The random displacements, $\Delta \mathbf{R}_\alpha$, of the atoms are coupled to the temperature in the classical simulated annealing scheme[117] according to a Gaussian distribution

$$P(\Delta \mathbf{R}_\alpha) \propto \exp\left(-(\Delta \mathbf{R}_\alpha)^2/T\right) \quad . \tag{7.7}$$

This coupling ensures that with decreasing temperature the step width is reduced more and more, thus effectively «freezing» the atom positions in the ground state configuration. To ensure convergence, the cooling rate by which the temperature is reduced within each iteration needs to be inverse logarithmic in time, i.e.

$$T \propto \frac{T_{start}}{\log(1+t)} \quad , \tag{7.8}$$

which has the drawback of quite slow convergence behavior. To improve the speed of convergence, adjustments to the annealing scheme can be made[118], in which the Gaussian distribution of displacements are replaced by Cauchy distributed displacements, thus allowing for larger jumps in the configurational space. With a position update like that, the logarithmic cooling rate can be replaced by a rate which is linear in time and therefore freezing the system much faster. Also adaptive procedures which adjust the cooling rate dynamically during the optimization run, e.g. *adaptive simulated annealing (ASA)*[119], are proven to work and increase the convergence speed.

By the successive reduction of the temperature due to the cooling rate no canonical ensemble is sampled anymore in the overall procedure. Therefore, no meaningful averages such as thermodynamic quantities can be extracted anymore, restricting the application of this procedure to pure optimization problems. However, as stated before, to extract meaning-

7.2. Structure Optimization by Global Optimization

ful ground state quantities, such as parameterized lateral interaction parameters, or structural information about ground state configurations, the importance of efficiently sampling as complete as possible the full configurational space cannot be underestimated.

7.2.2 Random Particle Placement Followed by Local Energy Minimization

A quite simple and yet successful approach to overcome the constraints of minimization techniques has been proposed by Pickard and Needs [120]. The idea is to place atoms randomly inside a simulation cell and then relax by a local energy minimization procedure into the next local minimum. Sine the starting configuration is random, there exists a high possibility of finding a large number of different local minimum structures for the given number of atoms to be distributed.

Moreover, as shown in Fig. 7.1 defining an active volume right above a surface allows to effectively search the configurational space of different adsorbate structures. By means of this stochastic sampling of the configurational space, however, a less controllable randomness compared to the simulated annealing algorithm is introduced. Moreover, there is no way to determine whether the sampling of the configurational space was exhaustive and the true global minimum has been found.

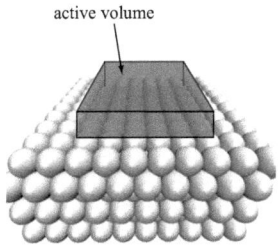

Figure 7.1: The definition of an active volume for the random placements of particles allow for an effective search for adsorbate cluster structures on a surface

The resulting local minimum structures are tabulated and sorted by total energy and the configuration lowest lying in energy is assumed to be the one closest to the global minimum. One indication, whether or not the sampling of the configurational space was exhaustive can be drawn from the observation that a certain structure is found much more often than others. Obviously, the possibility to find the real global minimum structure increases when only a few number of atoms are inside the active volume and this volume being not too large. In addition, one can influence the algorithm, by using some physical intuition in the definition of the active volume area, for example, by constraining the height of the active volume above the surface for a search of adsorbate structure which should consist of not more than a certain number of layers.

As for the simulated annealing, this algorithm does not rely on the specific method used for the energy evaluation, so that both methods are in principle applicable with first-principles energies. However, for more complex systems containing more but a few atoms, the energy evaluation of a large number of different structure on a first-principles level is computationally nowadays still unfeasible.

7.3 The Stability of Optimized Structures

Once the most stable minimum of the potential energy surface of all sampled configurations has been identified by search algorithms like the ones described before, the next question arising is naturally the one of the stability of this particular minimum configuration. This is the question for the long time dynamics of a thermal system (with fixed number of particles, i.e. the *canonical ensemble*). Real systems usually have a finite temperature, therefore they possess a certain thermal energy equal to $k_B \cdot T$. The stability of any (local or global) minimum in the thermodynamical limit is determined by the depth of the potential well in the PES that belongs to it.

A very nice discussion of this and the content of the following can be found in the excellent review article by Voter *et al.* [121].

7.3.1 The Minimum Energy Path and Transition Energy Barrier Definition

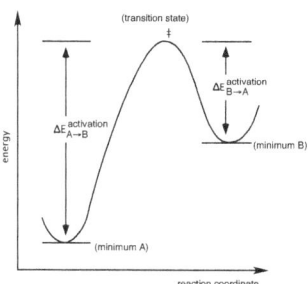

Figure 7.2: The definition of the activation barrier for a transition from a state A to another state B is shown here. It is defined as the energy difference $\Delta E^{\text{activation}}_{A \to B}$ between the energies of the lowest lying saddle point of the potential energy surface connecting the two local minima A and B with the minimum energy path (MEP) and the energy of the initial configuration.

Transitions from one minimum to another happen along a path orthogonal to the equipotential contours of a potential energy surface (PES) that connects the corresponding mimina through a saddle point. Along this path the force acting at any point is only pointing along that path. The barrier assigned to that transition is identified with the difference between the total energy at the saddle point and the total energy of the minimum.

$$\Delta E^{\text{activation}} = E^{\ddagger} - E^{\text{initial}}_A \qquad (7.9)$$

A transition between these two minima happens with a probability that is proportional to this energy barrier. The pathway along which the transition occurs, the connection between the initial and final state along the lowest lying path in energy is called the *minimum-energy path (MEP)* or the *minimum-energy reaction path (MERP)* as defined in the IUPAC Gold Book[122]. Therefore, determining the corresponding barrier of the transition between two different structures provides insight and allows assumptions about the stability of a configuration.

7.3.2 The Nudged Elastic Band Method

In the following subsection an algorithmic approach to the determination of the minimum energy path between two known structures and therefore the determination of the barrier for the transition will be presented, the nudged elastic band (NEB) method.[123] It is not the only one for finding transition states and there are many others, but throughout this study, this method was employed to get an estimate about the stability of the island structures presented in chapter 9.

In chain-of-states methods[124,125] (like the NEB) a string of geometric configurations of the system (usually called *images* in the context of NEB) is used to describe the reaction pathway for which the transition barrier should be determined. These images are connected via spring like forces to ensure equal spacing along the reaction path. Obviously, good knowledge about how the initial and final configuration look like is necessary.

The configurations along the spring connected chain of images are then relaxed in the NEB algorithm according to a force projection scheme, which takes into account two force components for each relaxation process of each image. One part of the forces is composed according to the curvature of the potential at the current configuration. The second component is calculated according to the image interconnecting spring forces introduced by the NEB scheme. The starting point for the projection is the definition of the tangent along the path, $\hat{\theta}$, as the unit vector to the higher energy neighboring image.[126] At extrema the vectors are interpolated via a linear interpolation scheme. Now the total NEB force on image i is given by

$$\mathbf{F}_i^{NEB} = {}^\perp \mathbf{F}_i^{pot} + {}^\parallel \mathbf{F}_i^{spring} \quad , \tag{7.10}$$

where ${}^\perp \mathbf{F}_i^{pot}$ is the perpendicular force acting on the tangent θ due to the curvature of the PES and ${}^\parallel \mathbf{F}_i^{spring}$ is the spring force parallel to the band of images. The latter are given for the ith image at position $\mathbf{R}_i$ and a spring with force constant k by

$$^\perp \mathbf{F}_i^{pot} = -\nabla(\mathbf{R}_i) + (\mathbf{R}_i) \nabla(\mathbf{R}_i) \cdot \hat{\tau}_i \hat{\tau}_i, \tag{7.11}$$

$$^\parallel \mathbf{F}_i^{spring} = k \left(|\mathbf{R}_{i+1} - \mathbf{R}_i| - |\mathbf{R}_i - \mathbf{R}_{i-1}| \right) \hat{\tau}_i, \tag{7.12}$$

After the calculation has converged, the pathway described by the images equals the true transition mechanism up to the resolution of the number of images. Typically, the image highest in energy is taken to represent the transition state and its energy is used for the calculation of the transition state energy. However, it must be stated clearly, that NEB is not a saddle point search algorithm and therefore it is not ensured that the highest energy image really equals the saddle point and therewith the reference energy for the transition energy barrier as defined above. Still, a lot insight can already be gained by the estimate for the energy barrier provided by the converged NEB pathway.

To converge to the true transition state, a small extension to the NEB algorithm has to be made, known as *climbing image nudged elastic band (CI-NEB)*. In this extended scheme, the force update for the image highest in energy is changed and the sign of the acting forces is reversed and all spring forces on this image are nullified. The resulting acting forces on this image are thus

$$\mathbf{F}_{E_{\max}}^{\text{CI-NEB}} = \mathbf{F}_{E_{\max}}^{\text{CI-NEB}} - 2 \cdot \mathbf{F}_{E_{\max}}^{\text{CI-NEB}} \cdot \hat{\theta}_i \hat{\theta}_i \quad . \tag{7.13}$$

In this way the coordinates of this configuration are moving uphill the curvature of the PES. This will converge to the point of the PES with highest energy and zero forces, which then equals the transition state configuration.

7.3.3 Transition State Theory

For the description of the dynamical evolution of a system and therefore the stability of certain configurations in the thermodynamic limit, the quantity that must be evaluated is not the barrier alone but the corresponding transition rate for the respective process.

In transition state theory (TST)[127–129], the classical rate constant for an escape from an initial state A to an adjacent state B is taken to be the equillibrium flux through the dividing surface between A and B of the PES as defined above. If no correlated events occur, this definition of the TST rate constant is the exact rate constant for this process. This definition equals the time average picture to extract the stability information of a given state introduced before. For illustrating purposed, imagine a two-state system and a very long classical direct MD trajectory. After a very long time, an equilibrium state will be established and both states will have been visited a large number of times. Now, one could determine the fraction of time, χ_A that the system spent in state A and the number of crossings, per unit time, of the dividing surface. The TST rate constant for the escape from state A would now be given by half of the crossing rate — accounting in this way only for escapes from A and not the events entering state A — divided by the time fraction χ_A.

This flux is thus an equilibrium property of the system (which can even be computed without ever propagating a real molecular dynamics (MD) trajectory by the appropriate statistical ensemble average — here, to account for temperature dependence it is the canonical ensemble)

$$k_{A\to}^{\text{TST}} = \left\langle \left| \frac{\mathrm{d}x_1}{\mathrm{d}t} \right| \delta(x_1 - q) \right\rangle_A \quad , \tag{7.14}$$

where the angular brackets are indicating the ratio of Boltzman-weighted integrals over $6N$-dimensional phase space (configurational space: $\mathbf{r}$, and momentum space: $\mathbf{p}$) restricted to the space belonging to state A — as indicated by the index. As a result of this restriction, there is no need to divide by the time fraction χ_A in this case anymore. The dividing surface

7.3. The Stability of Optimized Structures

is set to be at $x_1 = q$ for simplicity and the observation is restricted to the reaction coordinate x_1 only. x_1 is part of the configurational phase space of the system.

Calculating the average for some property $P(\mathbf{r},\mathbf{p})$ is then done by evaluating

$$\langle P \rangle = \frac{\iint P(\mathbf{r},\mathbf{p})\exp\left[-H(\mathbf{r},\mathbf{p})/k_B \cdot T\right]d\mathbf{r}d\mathbf{p}}{\iint \exp\left[-H(\mathbf{r},\mathbf{p})/k_B \cdot T\right]d\mathbf{r}d\mathbf{p}} \quad , \tag{7.15}$$

where k_B is the Boltzmann constant and $H(\mathbf{r},\mathbf{p})$ is the total energy of the system (kinetic plus potential).

If the effective mass of the reaction coordinate is constant while crossing the dividing surface, one can simplify equation 7.14 to a simpler average expression, averaging only over configurational space

$$k_{A\to}^{\text{TST}} = \sqrt{2k_B T/\pi m} \, \langle \delta(x_1 - q) \rangle_A \quad . \tag{7.16}$$

In this average the delta function picks out the probability of the system being at the dividing surface, relative to everywhere it can be in the basin of state A. There is no dependency of the nature of state B included whatsoever for this probability.

With this point of view it becomes possible to use this statistical framework to obtain accurate dynamical information about transition rates based on energy barriers. To actually follow the real dynamical evolution of the system in its full complexity is then not necessary anymore to obtain dynamical information.

Transition State Theory relies apart form the Born-Oppenheimer approximation on two basic assumptions:

- the rate is slow enough that a Boltzmann distribution is established and maintained in the reactant state;
- a dividing surface of dimensionality $D-1$, where D is the number of degrees of freedom in the system (usually $D = 3N$, with N being the number of particles), can be identified such that a reacting trajectory going from the initial state to the final state only crossed the dividing surface once. The dividing surface is therefore representing a bottleneck for the transition.

Harmonic Transition State Theory

For activated processes of the kind of systems that we are interested in here right now, i.e. crystal surfaces with adsorbates and cluster ordering, the harmonic approximation to TST can usually be applied[130,131], leading to *harmonic transition state theory* (hTST).

The rate constant for a transition can then accordingly be expressed by the energy difference between initial and transition state, see above, and a pre-factor calculated by the frequency of normal modes at the saddle point and the initial state[132,133]:

$$k^{\text{hTST}} = \frac{\prod_i^{3N}(v_i^{\text{init}})}{\prod_i^{3N-1}(v_i^{\ddagger})} \exp\left(\frac{-\Delta E^{\text{activation}}}{k_B \cdot T}\right) \tag{7.17}$$

where N gives the number of vibrational degrees of freedom. Within the harmonic approximation, where $h\nu \ll k_B T$, the vibrational partition function for the initial ($Z_{\text{vib}}^{\text{init}}$) and transition state ($Z_{\text{vib}}^{\ddagger}$) can be expressed by

$$Z_{\text{vib}}^{\text{init}} = \prod_{i=0}^{3N} \frac{k_B T}{h \cdot \nu_i^{\text{init}}} \quad \text{and} \quad Z_{\text{vib}}^{\ddagger} = \prod_{i=0}^{3N-1} \frac{k_B T}{h \cdot \nu_i^{\ddagger}} \quad . \tag{7.18}$$

These partition functions can then be used in equation 7.17 to yield an expression often seen in the literature

$$k^{\text{hTST}} = \frac{k_B T}{h} \frac{(Z_{\text{vib}}^{\ddagger})}{(Z_{\text{vib}}^{\text{init}})} \exp\left(\frac{-\Delta E^{\text{activation}}}{k_B \cdot T}\right) \quad . \tag{7.19}$$

If now every crossing of the dividing surface leads to a transition to the final state, the transition rate given by these equations is rather accurate. However, in reality, not every crossing leads to a transition and therefore, the transition rates of (h)TST usually overestimate the true rate constants.

7.3.4 Molecular Dynamics – Following the System's Dynamics Directly

Another way to gain information about the stability of certain configurations is to follow the time evolution of the system directly by treating the motion of the atoms using classical dynamics and performing a molecular dynamics (MD) simulation[134]. Within this approach the system will spend time in certain local minima according to their relative stability and the longest time in the infinite time limit will accordingly be spent in the most stable — global — minimum configuration. Consequently, the information about the stability of a certain configuration could be extracted by a statistical analysis of the MD trajectory, as already mentioned in the introduction to the transition state theory framework above.

However, the interesting transitions of most solid surface systems will be so-called *infrequent-events*[121]. The direct dynamical evolution of the system will consist of vibrations within the potential basin of the local/global minimum configuration and only occasionally a transition to another potential basin will occur. The average time between these transition events is typically very many vibrational periods, thus making the transitions infrequent with respect to the time scale of the system's vibrations. However, these vibrations have to be resolved in the time integration to maintain accuracy of the time evolution. The difference in time scales can easily be of the order of many orders of magnitude, so that a direct calculation is not feasible in cases, where the interesting long-time dynamics are characterized by extended residence times in the local minima of the PES. The frequency of occurrence of such a transition per unit time is termed *transition rate*.

When a transition happens, i.e. a trajectory moves from one potential basin of the $3N$ dimensional phase space to another minimum, this trajectory crosses a $3N - 1$ *dividing surface* at the ridge top separating the two minima. Even though these crossings are infrequent, successive crossings can happen on the time scale of just one or two vibrational periods; such

7.3. The Stability of Optimized Structures

transitions are then termed *correlated*. This defines the so-called *correlation time*, τ_{corr}, of the system, which can be understood as the system's memory. Once this correlation time has passed, the system does not know anymore how it got into the current state and the probability for a new transition from the the current configuration is independent from the way this configuration was reached. This means that the probability of escaping from a certain state is proportional to the transition rate constant for a certain transition path, which holds after the correlation time has passed for all states visited by a classical MD trajectory. This result lays the basis for a framework to seize the independence of escape events for an acceleration of direct MD simulations. Normally, these simulations are intrinsically serial in time, however, the next session will describe a technique for parallelization based on a parallel exploration of the different escape paths of a potential well.

Accelerated Molecular Dynamics – Parallel Replica Dynamics

As one of the simplest and most accurate schemes to accelerate direct molecular dynamics simulations employing the principles of transition state theory, the method of *parallel replica dynamics*[135] (PRD) represents in particular for systems studied at quite elevated temperatures[121] a very appealing choice for an acceleration scheme.

The only assumption underlying the formalism of parallel replica dynamics is that the infrequent events of the system under study obey a first-order kinetics (exponential decay). This is, for any time t greater than the system specific correlation time (see above), τ_{tau}, after entering a state A, the probability distribution function for the time of the next escaping transition from that state is given by

$$p(t) = k_{A\rightarrow} \exp(-k_{A\rightarrow} \cdot t) \quad , \qquad (7.20)$$

where $k_{A\rightarrow}$ is the transition rate constant for escape.

The general approach is shown in figure Fig. 7.4. A system residing in a potential basin A is replicated identically on each of M available CPUs. After a short phase of de-phasing for a time $t_{dephase} \geq \tau_{corr}$, during which the particle momenta are randomized to eliminate correlation and which is performed long enough to be greater than the correlation time, each processor carries out an independent constant-temperature direct MD trajectory for the system replicated on it. The specific correlation time that needs to be used depends on the specific system under investigation, typically for diffusion processes on solid surfaces it ranges in the order of 1 ps[135]. Its specific value for the system under investigation can be extracted from a time-correlation analysis of an normal MD simulation for this system[130].

In periodic intervals on all of the parallel MD simulations a local structure optimization (see above, section 7.1) is performed to check for a transition event. A transition is defined to have happened, if the local energy minimization leads to a local minimum structure that is distinguishable from the initial starting structure. Usually the criterion for this are the atomic positions of the system that is simulated.

If a transition is found on one of the replicas, all processors are alerted to stop. The global simulation clock, t_{global} is advanced by the accumulated trajectory time of all replicas, i.e.

Chapter 7. System Optimization

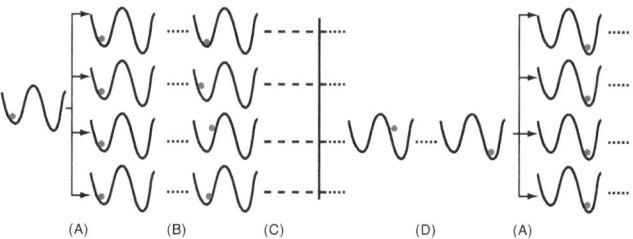

Figure 7.4: In this figure, the parallel replica dynamics method is schematically illustrated. Starting from a given configuration, in step (A) a certain number of identical replicas are generated. After a dephasing period (B), independent MD trajectories are run for each replica of the system at (C). Once a transition is found, this specific replica is soley run for a certain correlation time in (D). After this, the replica where the transition happened is the starting configuration for the next PRD run and the total simulation time is the sum of all trajectory times of all replicas.

the total time spent from random starting configurations to explore the local energy basin of state A before an escape transition happened. After this, the one replica for which the transition event occured is allowed to follow its trajectory for a time Δt_{corr} longer than the correlation time for that system, $\Delta t_{\text{corr}} \geq \tau_{\text{corr}}$ to check for correlated events and/or recrossing of the dividing surface. After this, the global clock is advanced by this Δt_{corr} and the final state of the replica after the transition event is the new starting state that is copied to all available processors. After this, the whole loop starts again.

For the escape-time probability of the super-system consisting of the M replicas (and assuming for simplicity that all processors are running at the same clock-speed, so that the summed speed $S = M$) as a function of t_1, the trajectory time on processor 1, we can write

$$p_{\text{super}}(t_1) = S \cdot k_{A\rightarrow} \exp(-S \cdot k_{A\rightarrow} \cdot t_1) \quad . \tag{7.21}$$

Now, at a given point in time, the accumulated simulation time summed over all M trajectories is related to t_1 by

$$t_{\text{sum}} = S \cdot t_1 \quad . \tag{7.22}$$

Inserting this into equation Eq. 7.21 and recognizing that for all probability distributions p the relation

$$\frac{1}{S} p\left(\frac{t}{S}\right) dt = p(t) dt \tag{7.23}$$

holds, gives the relation

$$p_{\text{super}}(t_{\text{sum}}) = k_{A\rightarrow} \exp(t_{\text{sum}} \cdot k_{A\rightarrow}) \quad . \tag{7.24}$$

7.3. The Stability of Optimized Structures

Therefore, running M independent replica simulations and defining the summed time according to equation Eq. 7.22 shows, comparing to the fundamental equation Eq. 7.20, that this procedure generates the *correct probability distribution for the escape time for the initial state*. This is also valid for replicas run on processors of different speed[135].

The parallelization has no effect on the relative probabilities for the possible escape paths. This means, if the parallel running replicas are continuously monitored for transition events, the resulting sequence of events for the parallel run is indistinguishable from a sequence gained by a serial run MD trajectory. However, in the procedure described above an exact mapping between the system's configuration and the total simulation clock can only be made at the time of each transition event, or w.r.t. a parameter for the simulation to set, on the time scale of the intervall in which all replicas are examined for a transition event. Finer time scales can be obtained from any of the individual replicas, however correlating them to the total clock is not possible.

Part III
Application and Results

Chapter 8
The Adsorption of CO on the Pd(100) Surface

ONE can maybe say that the adsorption of CO on single crystal surfaces, in particular (100) facets of simple bcc and fcc metals is one of the most investigated adsorption reaction. The reasons for this are naturally its fundamental importance as a model system for heterogeneous catalysis applications. The basic principles of metal–molecule interactions and principles of chemisorption can be studied using this well-defined system of still controllable complexity.

As this system is representing a showcase for the case, where a strong bond to the substrate is formed and the interaction amongst the adsorbed particles are rather weak compared to the surface–adsorbate bond, this system has been examined theoretically as the example of this thesis for the second limiting case of the Frenkel-Kontorova model (as introduced in the introduction, see chapter 2).

8.1 The Pd(100) Surface and its Adsorption Sites

The simple high symmetry (100) surface of the fcc structure of Pd is schematically shown in Fig. 8.1 to illustrate the basic geometric features of this surface facet. The surface unit cell can be given in two different ways, both of which are also shown. The primitive (1 × 1) surface unit cell is the simplest possible one, containing the minimum number of atoms, whereas the centered $c(2 \times 2)$ contains two atoms in the unit cell, but still exhibits all symmetry features of the highest surface point group, C_{4v}. Since all experimentally found adsorption structures of CO on this surface arrange in periodic structures based on this $c(2 \times 2)$ cell, cf. Fig. 8.2, all calculations and descriptions of adsorbate structures within this thesis will be given in terms of the centered surface unit cell. However, since usually the coverage, θ, is given with respect to the one hollow position enclosed in the $p(1 \times 1)$ unit cell, the latter will represent the reference structure for all coverage values in this study. Consequently, a $\theta = 1$ ML (ML = mono layer) coverage for the $c(2 \times 2)$ unit cell is reached when 2 of the available 4 bridge

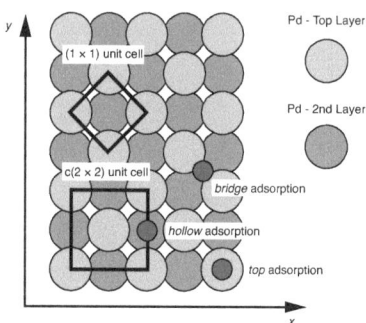

Figure 8.1: Schematic view of the (100) surface of the face-centered cubic structure along the cristallographic z-axis. Shown are the two top layers of the surface, light grey circles represent the top layer atoms, darker grey circles are representing the atoms of the second layer from the top. Two different surface unit cells are shown, the primitive (1×1) cell and the centered $c(2 \times 2)$ cell.
The red circles in the structure depict the positions of the three distinct high symmetry adsorption sites of this facet of a fcc metal surface, namely the *hollow* site, the *bridge* site and the *top* site.

adsorption sites are occupied with adsorbed CO molecules.

Shown as red circles in Fig. 8.1 are the three different high-symmetry adsorption sites of this surface. The four-fold hollow site, the bridge site and the on-top adsorption site. In the primitive (1×1) surface unit cell one site of each kind is included within the unit cell, whereas the bigger $c(1 \times 2)$ cell contains two of each kind of adsorption site. Depending on the kind of adsorbing species and the physics behind the interactions between surface and adsorbate, either of those sites can be the most favorable adsorption site for a single adsorbed particle.

8.2 CO Adsorption - Experiments and Theory

As already mentioned, the CO adsorption on Pd(100) represents a showcase system to experimentalists for studies of adsorption. Already in 1969 this system was used by Tracy and Palmberg for the investigation of chemisorption and as a showcase for the determination of binding energies[136]. In a later intensive study, Bradshaw and Hoffmann combined infrared adsorption spectroscopy (IRAS) and low-energy electron diffraction methods to investigate the detailed adsorption site of the adsorbate on the surface[137], followed by a work of Behm et al., which characterized thoroughly the $\theta = 0.5$ ML configuration and addressed in particular for the first time the bonding geometry of the CO molecule at the adsorption site[138,139]. In all those studies, CO was found to adsorb upright in the bridge position of the underlying Pd(100) surface. Further LEED[140] and IRAS[141,142] studies confirmed the ordered structures and are in line with the results of surface core level shifts (SCLS)[143].

8.2.1 Ordered Overlayer Structures of CO on Pd(100)

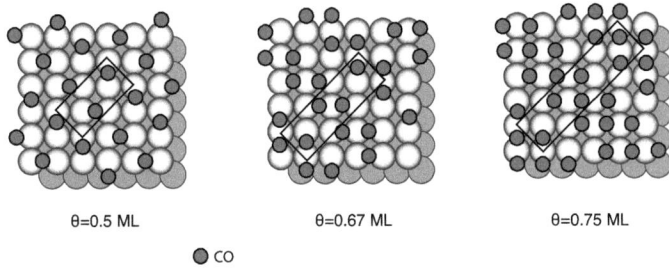

θ=0.5 ML   θ=0.67 ML   θ=0.75 ML

○ CO

Figure 8.2: Shown here are the three experimentally characterized ordered overlayer structures for CO adsorbed on the Pd(100) surface for coverages of $\theta = 0.5$ ML, $\theta = 0.67$ ML and $\theta = 0.75$ ML coverage. All coverages are given with respect to the simple primitive (1 × 1) surface unit cell, which exhibits one hollow, one bridge and one top sites, whereas for the bigger $c(2 \times 2)$ unit cell there are two sites of each kind.
The tendency of CO to try to arrange in a fashion that keeps the distance between CO molecules maximal has been attributed to strongly repulsive lateral interactions on this surface.

The presented ordered overlayer structures of CO begin to form at a coverage of $\theta = 0.5$ ML. At this value an ordered $(2\sqrt{2} \times \sqrt{2})R45°$ structure is formed. This structure is compressed to a $(3\sqrt{2} \times \sqrt{2})R45°$ structure when a coverage value of $\theta = 0.67$ ML is reached. The highest coverage for which an experimentally characterized structure is known is $\theta = 0.75$ ML. This structure is represented by a $(3\sqrt{2} \times \sqrt{2})R45°$ unit cell. All these structures are shown in Fig. 8.2.

The formation of a $(2\sqrt{2} \times \sqrt{2})R45°$ structure rather than a simple $c(2 \times 2)$ structure at $\theta = 0.5$ ML is usually interpreted as a tendency of the overlayer system to minimize the strongly repulsive interaction between the neighboring CO molecules. In the realized ordered structure every CO molecule has two neighbors within a distance of $a_{100}\sqrt{2}$ and four neighbors in a distance of $a_{100}\sqrt{5/2}$, whereas the $c(2 \times 2)$ configuration results in four neighbors in the distance of $a_{100}\sqrt{2}$, therefore on average a slightly closer packing of the adsorbates on the surface.

The behavior of the CO overlayer in these high coverage regimes has also been quite intensively studied. In particular, the behaviour of the ordered structures at finite temperatures is a matter of current debate, whereas the three structures presented in Fig. 8.2 are generally accepted as the low temperature ordered structures[139,140,144,145]. Not only the (100) surface is a prominent example system, but also the adsorption of CO on the (111) facet of palladium is taken as a showcase[146]. The influence of CO partial pressure on the adsorption behavior on both facets has been studied[147] and intermixing of adsorbed particles in co-adsorption experiments was investigated as well[148].

The origins of the debates about dense adlayer structures are to be found in the fact that the experimental techniques (work function measurements and infrared analysis) are facing some difficulties for high coverages with (θ > 0.5 ML). In addition, the appearance of a phase transition to an incommensurate overlayer structure studied in diffraction experiments at temperatures T > 350 K has been observed already for coverages of θ = 0.5 ML[149]. This observation would imply that at higher temperatures and for dense coverages the ordered structures shown in Fig. 8.2 might not be stable anymore.

Turning from structural stability to interaction parameters, nowadays there are attempts to assess the interactions between adsorbates also more directly. One example for this is by means of time resolved STM experiments. This has been done also for the system CO/Pd(111)[150]. As a result, repulsive interactions were concluded from the measurements for the first and second nearest neighbor interaction terms, whereas for a third nearest neighbor interaction on this surface, a very weak attraction was proposed.

8.2.2 Theoretical Work

Adsorption Energetics and Static DFT Calculations

Previous theoretical work is mostly focused on the adsorption energetics and the relative stability of different adsorption sites[151–153]. Contrary to the case of platinum[154,155], however, DFT using the GGA-PBE xc-functional predicts the correct stability order of the different high symmetry adsorption sites in agreement with experiments (bridge > hollow > top) for palladium. Dense configurations of adsorbed CO molecules on this surface are mainly included as static configurations, since tackling the interplay of large quantities of adsorbed molecules is theoretically quite challenging, if one wants to explore the configurational space extensively. Also the interplay of adsorbed CO molecules with other adsorbed species is predominantly theoretically studied only within a static DFT picture[156]. The Blyholder model[157,158] for the CO adsorption on metal surfaces was one of the first theoretical models that tried to describe the process of chemisorption on a electronic level/molecular orbital level and it still provides some insight into the physical interactions underlying the adsorption process.

As already mentioned in the introduction to the embedded-atom-method, a DFT calculation intrinsically includes all static pair and many body interaction in the calculated total energy values, however the theoretical work based on DFT calculations, did not focus on the extraction of the relevant parameters to describe the interaction between adsorbates.

Previous Work about the Interactions between Adsorbed Molecules

The adsorbed CO molecules can be seen in a first schematic approach as an array of parallel oriented dipoles. Following this line of modeling the influence of lateral interactions on a measurable quantity, namely the vibration spectrum, was done by Scheffler in 1979[7]. Here the observed coverage dependency of the vibrational signal was explained by a simple picture of interacting dipoles. The interactions taken into account were restricted to a pair

8.2. CO Adsorption - Experiments and Theory

interaction picture between neighboring particles and the focus of the theoretical description was more laid on the effect of the oscillating IR field on the array of dipoles, rather than the extraction of interaction parameters.

Statistical approaches which are trying to cover also the dynamics of the interplay between the adsorbed molecules on the surface and the influence of this interplay on ordering or reactivity are also present and mainly rely on a cluster expansion like approach for the configurational energy of different configurations.[159–163]. However, all cited theoretical work relies on the approach of fitting a few interaction parameters (typically some pair and sometimes a very limited number of selected trio interactions) of the theoretical model until this model reproduces experimentally observed configurations to satisfying accuracy. Also some very recent work[163] follows the same fitting-to-experiments based approach.

There are clearly two systematic drawbacks in such a theoretical approach. First, those interaction parameters derived from fitting with the aim to reproduce macroscopic configurations do not necessarily have any physical meaning on the microscopic level. Secondly, the very limited number of interaction parameters included in the fit set in these simulations and the lack of a consistent procedure to select the important parameters presents a major deficiency.

For comparison, however, such reported parameters are representing a good starting point. Liu and Evans reported values for the next-nearest neighbor interaction (V_1^p in this work) of infinite repulsion (site exclusion), for the second next-neighbor interaction (V_2^p) a value of +170 meV (where a plus sign indicates repulsion) and finally for the third next-neighbor interaction (V_3^p) a weak repulsion of +2 meV[161].

8.3 First-Principles Lateral Interactions of CO on Pd(100)

In this part of the thesis, a systematic approach to the determination of lateral interaction parameters presented in chapter 5, the cluster expansion formalism, will be used in an attempt to derive a consistent set of lateral interaction parameters for the CO adsorption on the Pd(100) surface from first-principles.

8.3.1 The Pair Interaction of a Parallel CO Dimer

As an illustrative example about the dimension and kind of interaction between CO molecules aligned parallel to each other, similar to the configuration after being adsorbed on a surface, Fig. 8.3 shows the repulsive interaction of a dimer of CO molecules in vacuum, calculated at the GGA-PBE DFT level with the parameter setup used for the gas-phase reference state calculations described in appendix A. From this, one can get already a first confirmation about the fact that the pure pair interaction can be expected to

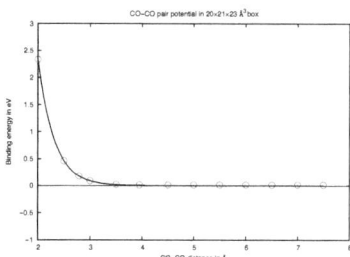

Figure 8.3: The pure pair interactions of a dimer of CO molecules oriented parallel to each other in vacuum. Shown are as red circles the calculated data points, the line is a fitted Morse potential for illustrative purposes/.

be rather strongly repulsive, even taking into account that the shown vacuum repulsion is definitively lacking the expected screening effects due to the interaction with the surface.

8.3.2 The On-Site Binding Energy

A second quantity of interest is presented by the pure bond formation energy of a CO molecule to its favored bridge adsorption site of the Pd(100) surface without any contributions from interacting molecules. To extract this value, a convergence test like series of calculations have been performed in which a single CO molecule was adsorbed in a simulation unit cell of three different sizes, leading to a larger and larger separation of periodic images.

The results for the calculations within unit cells of size $c(2 \times 2)$ with a separation of 3.95 Å, of size $c(3 \times 3)$ with a separation of 7.90 Å and of size $c(4 \times 4)$ with a separation of 11.85 Å between neighboring CO molecules are shown in Fig. 8.4. The final value for the largest separation, $\Delta E_{bond} = -1957.3$ meV, has been taken as the *on-site energy* term in the cluster expansion later on. Visible in this plot is the effect of the long range repulsive interaction caused by the interacting dipoles of the adsorbed CO molecules. Even if there is a clear tendency visible for a further increase in the on-site bonding energy, it is also visible that

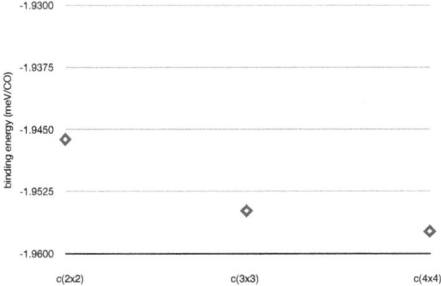

Figure 8.4: Binding energy of one CO molecule in meV for the adsorption at a bridge adsorption site for three different unit cell dimensions.

the bonding energy is converged within ≤ 3 meV for the last step from the $c(3 \times 3)$ to the $c(4 \times 4)$ unit cell. This uncertainty is far below the accuracy that we are aiming for in the cluster expansion approach. For this reason, no further enlarged unit cell sizes were taken into account.

8.3.3 Direct Determination of Pair Interactions from DFT calculations

A direct usage of the information gained from the tests of the subsection before can be made in setting up two simple DFT configurations which will allow the direct determination of the first and second pair interaction parameters, V_1^p and V_2^p.

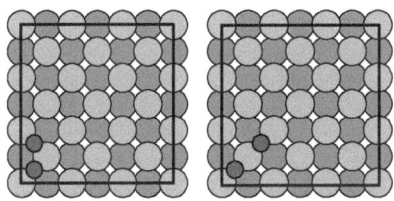

Figure 8.5: The two unit cell configurations used for the direct determination of V_1^p and V_1^p from DFT calculations.

The idea is that by using large enough unit cell sizes (in the case here these are presented by a (4×4) unit cell size in the setup) in the periodic DFT calculation, the influence of the coupling of periodic images is negligible and therefore, the binding energy in these calculations is only determined by the direct lateral interaction of the two CO atoms placed in neighboring adsorption sites. The corresponding adsorbate configurations in the simulation unit cell setup for the determination of the V_1^p and V_2^p interaction strengths directly from the resulting total DFT energy are presented in Fig. 8.5. The total binding energy in these configurations is given by the very simple CE:

$$E_{\text{bind}}^{\text{total}} = 2 \cdot E_{\text{bond}} + 2 \cdot V_x^p \quad , \quad (8.1)$$

where x is 1 in case of the left side of Fig. 8.5 and 2 in case of the situation depicted on the right side. The value of E_{bond} is hereby set to be the *on-site binding energy* in the (4 × 4) surface unit cell, as described in the subsection before.

By this simple procedure, the values for the interaction parameters were then calculated to be:

- $V_1^p = +188.6$ meV
- $V_1^p = +113.2$ meV

Therefore, the two simplest pair interactions are both strongly repulsive. This is in good qualitative agreement of the values reported by Liu and Evans before[161] and also in line with the experimental observations for the Pd(111) facet[150]. Our finite value for the first pair interaction is significantly larger than the value for the second pair interaction, which corresponds well to the fact that Liu and Evans used a site-exclusion rule within their model approach to reproduce experimentally observed order patterns of adsorbed CO. The second pair interaction term that we determined is slightly lower compared to the one of Liu and Evans +170 meV vs. +113.2 meV but still in remarkable agreement with their value. This holds in particular when noting that the reported parameters used in the model approach by Liu and Evans have been derived within a fitting process to reproduce the experimentally characterized ordered structures for a coverage of $\theta = 0.5$ ML, whereas our parameters are directly derived from first-principles calculations and do not include any fitting to experimental data.

As a consequence of this good agreement, the values for the first- and second-nearest pair interaction parameters will be taken as an additional benchmarking information next to the pure cross-validation score of any further cluster expansion in a set of lateral interaction parameters for a set of many configurations. Assuming that the direct determination of these values presented here is accurate, the values for these interaction parameters should be quite stable in any meaningful further cluster expansion and therefore remain almost constant at their directly determined values.

8.3.4 Coverage Dependent Binding Energy

In order to get a first idea about the effects that the interaction of the adsorbed molecules will have on the binding properties of the molecule to the surface the binding energy of several different structures normalized to one CO molecule is plotted in Fig. 8.6 as a function of the

8.3. The Lateral Interactions of Adsorbed CO on Pd

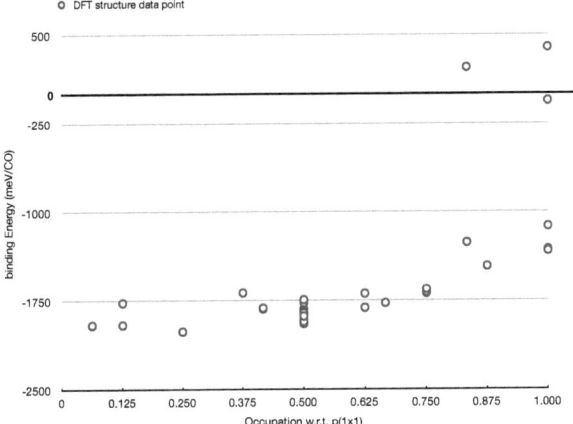

Figure 8.6: Binding energy per CO molecule given in meV for the adsorption on a bridge adsorption site plotted against the respective coverage (with respect to the $p(1 \times 1)$ surface unit cell hollow sites) for several adsorbate configurations calculated at GGA-PBE level.

coverage (given with respect to the $p(1 \times 1)$ hollow site). It can be seen that indeed the binding energy reduces with increasing coverage. Furthermore, very dense packing of CO only on the bridge sites seems to get more and more unfavorable until finally at the monolayer coverage some configurations calculated with locally very dense packing of CO molecules are becoming unstable with respect to desorption (these are the cases with positive binding energies).

Another observation is that (at least for the considered set of adsorbate configurations, which of course included all experimentally characterized structures) until a coverage of $\theta = 0.5$ ML the binding energies per CO molecule are scattered in a binding energy range window between -1950 and -1800 meV/CO but do not increase significantly. Above half monolayer coverage the reduction, however, is more pronounced. Obviously, this observation can not be generalized since for such an argument all possible configurations of adsorbates in all possible unit cell sizes and shapes for a given coverage should be evaluated within a direct enumeration approach.

8.3.5 Definition of the Set of Lateral Interaction Parameters

Next, the pool of interaction parameters for the description of lateral interactions needs to be defined. The total number of interaction parameters that can be considered in a cluster expansion is of course limited by the number of DFT input structures. However, the set of interaction parameters should be chosen such that it provides enough configurational flexibility to describe a lot of different adsorbate configurations to satisfying accuracy.

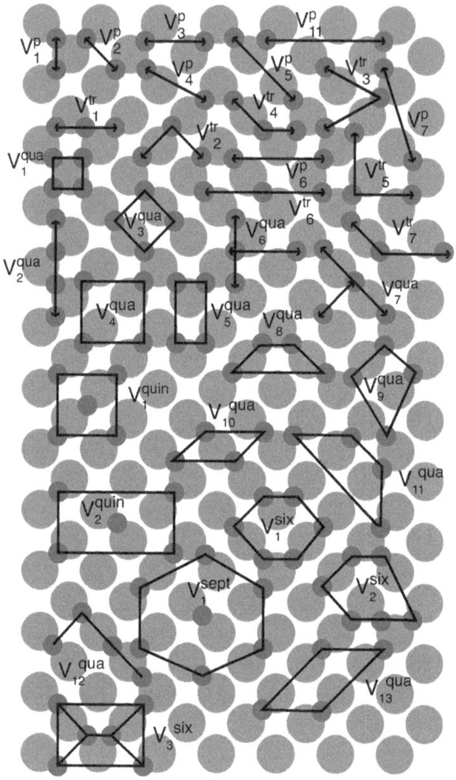

Figure 8.7: Shown here is the complete set of 34 lateral interaction parameters used as a basis for the cluster expansions of this work. Red dots indicate CO molecules adsorbed at bridge sites of the (100) surface. The corresponding interaction figure is marked with black lines and its name is given, where the superscript index indicates the type of interaction parameter and the subscript index is a counter.

Shown in Fig. 8.7 is the full set of 34 interaction parameters for the CO adsorption on the bridge sites of the Pd(100) surface considered during the process of determining a cluster expansion description of the interactions.

This set consisted of 8 pair interactions, whereas the very long range simple pair interaction V_{11}^p was mainly taken as an indicator whether a fitted set of parameters would consist

8.3. The Lateral Interactions of Adsorbed CO on Pd

of a physically meaningful sequence of parameter energies or not. The interaction strength should decay with distance and this should be clearly resembled by the sequence of interaction energies assigned to the pair interactions within the parameterization process. For this reason, for a set of fitted interaction parameters, the value for the long range pair interaction was checked, but for the later evaluation, this long range pair interaction was not included in the fit base anymore. In all reported cases, this lead only to a negligible increase of the CV score.

Seven different trio interactions, V_x^{tr}, are also included in this initial set, in addition to a set of 13 different quattro interactions, V_x^{qua}, two quinto interactions, V_x^{quin}, a set of three six-particle interactions, V_x^{six}, and finally one many-body interaction figure constituted by seven adsorbate atoms, V_1^{sept}.

8.3.6 Finding the Data Set of DFT Structures for the Parameterization

In addition to the already presented two $c(4 \times 4)$ structures for the direct determination of the pair interactions, initially 45 configurations were set up for the generation of the DFT database. These obviously included the three experimentally described structures for the coverages $\theta = 0.5$ ML, $\theta = 0.6667$ ML and $\theta = 0.75$ ML. The set of DFT structures included unit cell shapes of $c(2 \times 2)$, $c(3 \times 2)$, $c(4 \times 2)$, $c(5 \times 2)$, $c(6 \times 2)$, $c(3 \times 3)$, $c(4 \times 3)$ and as already mentioned $c(4 \times 4)$.

Distortion and Relaxation Effects for Structures with Local High Coverage

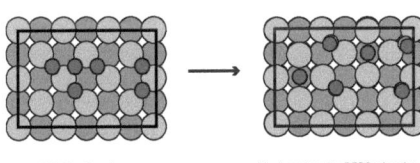

Figure 8.8: Example of major relaxation effects for locally dense adsorbate configurations for an example $c(4 \times 3)$ structure. In this example two CO atoms moved to hollow adsorption sites to avoid strongly repulsive interactions with neighboring CO atoms.

In the first set of DFT configurations a lot of configurations with locally very dense packing of adsorbates were set up to generate a database for the parameterization of the close many-body interactions. These particular structures, however, turned out to be unstable against relaxation already on the level of a local geometry optimization. In quite a number of cases, this local geometry optimization led even to configurations in which the CO molecules had moved from the initial adsorption sites to the more unfavorable hollow or even top sites of the underlying Pd(100) lattice. An example for these relaxation effects caused by very strong repulsive interactions is shown in Fig. 8.8. Here in a $c(4 \times 3)$ unit cell, the T-shaped configuration of adsorbed CO molecules dissolved upon geometry relaxation on the level of GGA-PBE to the configuration shown on the right side of the inset figure. Two CO molecules were pushed to the less favorable hollow adsorption site in the unit cell. This pushing and the following stabilization of this structure is caused by the strong repulsive interactions between the densely packed CO molecules in the initial configuration. For relaxations like this a fundamental assumption for the applicability of the cluster expansion method for the description of lateral interactions naturally breaks down. Switches of adsorption sites cannot be covered by the cluster expansion approach obviously, since the interaction parameters are statically defined for fixed positions of adsorbates.

This is also the reason for a second source of problems for the parameterization. The fact that adsorbed CO molecules in close proximity tend to react to the strongly repulsive interactions between them with strong axial bending of the molecule axis away from the ideal upright adsorption geometry, see Fig. 8.9 for an example, influences of course the value that would be assigned to this particular interaction figure within the fitting process based on the total energy of this particular configuration. The parameterization of this interaction figure would, however, only take into account the position on the surface and not the angle between the molecular axis and the surface normal. Another occurrence of the same figure but with different angle between the adsorbates will contribute with a different energy value for this interaction to the fitting database.

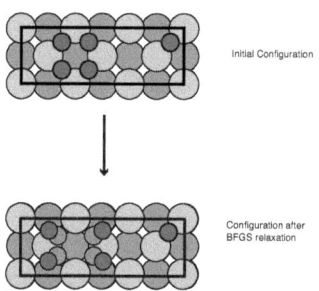

Figure 8.9: Shown here are the strong axial distortion effects upon geometry relaxation for locally dense CO adsorbate configuration. Although the initial adsorption site was kept during the relaxation, the angle between the surface and the molecular axis changed quite significantly, as visible in the lower part of the figure.

Consequently, these kind of effects will obviously influence the predictive power of the resulting cluster expansion and introduce uncertainties into the predicted CE binding energies for configurations that include an adsorbate arrangement like this.

Therefore, based on the relaxed geometries of the initial set of 47 DFT configurations, the ones which were heavily distorted and the ones for which upon geometry relaxation the initial bridge adsorption site was not occupied anymore (because the adsorbed CO was pushed to a different adsorption site) were sorted out from the set of input structures for the parameterization procedure of the cluster expansion. An additional set of 11 adsorption configurations with less dense packing of adsorbates in close proximity in a $c(4 \times 3)$ unit cell were then prepared in a next step and these replaced the structures sorted out before. Appendix B shows an overview over all the configurations calculated at the level of GGA-PBE. For each configuration a picture of the initial configuration and of the resulting configuration

8.3. The Lateral Interactions of Adsorbed CO on Pd

after the BFGS geometry relaxation is given to present the effects of local relaxations.

The Final Set of DFT Structures

The final set of adsorbate configurations with DFT total energies consisted of 37 different structures for the parameterization of maximal 34 interaction parameters (in case of the full-set fit). In addition to that the *on-site* bonding energy was separately parameterized as described above within a $c(4 \times 4)$ unit cell with only one CO adsorbed on a bridge site. The coverage range included within this set of DFT structures span coverages from $\theta = 0.05$ ML to a monolayer coverage (normalized as usual to the number of hollow adsorption sites in the $p(1 \times 1)$ unit cell). The fit was not improved terms of a lower cross-validation score by inclusion of the very densely packed (and energetically very unfavorable) formally 2 ML configuration, where all bridge sites surrounding one surface atom were occupied.

In detail, two $c(2 \times 2)$ structures with one and two CO adsorbed, 9 $c(3 \times 2)$ structures with the number of CO adsorbed ranging from one to four, three different adsorbate configurations in a $c(3 \times 3)$ unit cell with one, four or five adsorbed CO molecules were included in the set of structures. In addition, three $c(4 \times 2)$ and 12 $c(4 \times 3)$ structures were included with the number of adsorbed CO molecules ranging from two to five. The fit set was completed by the two $c(4 \times 4)$ structures with two adsorbed molecules, six $c(5 \times 2)$ structures with four to six CO molecules and one $c(6 \times 2)$ structure with nine molecules adsorbed. Pictures of the relaxed adsorbate configurations can be found in Appendix B and the corresponding total energies are given in table 8.1 below. The structures selected for the parameterization are marked with black bullets in this overview.

Identifier	coverage wrt to $p(1 \times 1)$ hollow	binding energy	formation energy in meV/CO	included in fit set
$c(2 \times 2)$ 2CO A	1.0	−1316.98	0.00	•
1 ML reference structure for the formation energy evaluation				
$c(2 \times 2)$ 1CO	0.5	−1966.16	−314.59	•
$c(2 \times 2)$ 2CO B	1.0	+390.49	+1707.45	
$c(2 \times 2)$ 3CO	1.5	+480.17	+2695.71	
$c(2 \times 2)$ 4CO	2.0	2419.71	+7479.36	
$c(3 \times 2)$ 2CO exp	0.5	−1942.83	−312.93	•
$c(3 \times 2)$ 1CO	0.25	−2010.73	−173.44	•
$c(3 \times 2)$ 2CO A	0.5	−1821.33	−252.17	•
$c(3 \times 2)$ 2CO B	0.5	−1820.37	−251.70	•
$c(3 \times 2)$ 2CO C	0.5	−1988.13	−335.57	•
$c(3 \times 2)$ 3CO A	0.75	−1689.25	−279.21	•
$c(3 \times 2)$ 3CO B	0.75	−1688.95	−278.98	•
Continued on next page ...				

Identifier	coverage wrt to $p(1 \times 1)$ hollow	binding energy in meV/CO	formation energy	included in fit set
\...continued from previous page				
$c(3 \times 2)$ 3CO C	0.75	−1626.37	−232.04	
$c(3 \times 2)$ 4CO A	1.0	−1330.54	−13.56	•
$c(3 \times 2)$ 4CO B	1.0	−842.28	+474.67	•
$c(3 \times 2)$ 4CO C	1.0	−1330.47	−13.49	
$c(3 \times 2)$ 4CO D	1.0	−1111.29	+205.68	
$c(3 \times 2)$ 5CO	1.25	−770.86	+682.633	
$c(3 \times 3)$ 1CO	0.125	−1954.82	−79.73	•
$c(3 \times 3)$ 3CO A	0.375	−1684.50	−137.82	
$c(3 \times 3)$ 4CO A	0.5	−1946.32	−314.67	•
$c(3 \times 3)$ 4CO B	0.5	−1770.75	−226.89	
$c(3 \times 3)$ 4CO C	0.5	−1767.28	−225.15	
$c(3 \times 3)$ 4CO D	0.5	−1745.02	−214.02	
$c(3 \times 3)$ 5CO A	0.625	−1692.92	−234.97	
$c(3 \times 3)$ 5CO B	0.625	−1790.65	−296.05	•
$c(3 \times 3)$ 6CO	0.75	−1652.15	−251.39	
$c(4 \times 2)$ 4CO exp	$0.6\bar{6}$	−1772.48	−303.67	•
$c(4 \times 2)$ 5CO A	$0.8\bar{3}$	−1258.54	+48.69	
$c(4 \times 2)$ 5CO B	$0.8\bar{3}$	+222.25	+1282.68	•
$c(4 \times 2)$ 5CO C	$0.8\bar{3}$	−1564.46	−206.23	
$c(4 \times 2)$ 6CO	1.0	−59.71	+1257.26	•
$c(4 \times 2)$ 7CO	$1.1\bar{6}$	−688.96	+732.68	
$c(4 \times 2)$ 8CO A	$1.\bar{3}$	−286.57	+1373.87	
$c(4 \times 2)$ 8CO B	$1.\bar{3}$	−606.26	+947.62	
$c(4 \times 2)$ 9CO	1.5	−775.92	+811.57	
$c(4 \times 3)$ 5CO A	$0.41\bar{6}$	−1821.60	−210.26	
$c(4 \times 3)$ 5CO B	$0.41\bar{6}$	−1814.53	−207.32	
$c(4 \times 3)$ 6CO A	0.5	−1908.22	−295.62	
$c(4 \times 3)$ 6CO B	0.5	−1929.08	−306.05	
$c(4 \times 3)$ 6CO C	0.5	−1874.22	−278.62	
$c(4 \times 3)$ 6CO D	0.5	−1844.08	−263.55	
$c(4 \times 3)$ 8CO A	$0.6\bar{6}$	−1748.81	−287.89	
$c(4 \times 4)$ 1CO	$0.0\bar{5}$	−1957.30	−35.57	
$c(4 \times 4)$ 2CO	$0.\bar{1}$	−1768.69	−50.19	•
$c(4 \times 4)$ 2CO A	0.5	−1844.09	−58.57	•
Continued on next page\...				

8.3. The Lateral Interactions of Adsorbed CO on Pd

Identifier	coverage wrt to $p(1 \times 1)$ hollow	binding energy in meV/CO	formation energy	included in fit set
...continued from previous page				
$c(5 \times 2)$ 6CO exp	0.75	−1678.87	−271.42	•
$c(5 \times 2)$ 4CO A	0.5	−1855.62	−269.32	•
$c(5 \times 2)$ 4CO B	0.5	−1881.89	−282.46	•
$c(5 \times 2)$ 5CO A	0.5	−1812.13	−309.48	•
$c(5 \times 2)$ 6CO A	0.75	−1654.77	−253.34	•
$c(5 \times 2)$ 7CO A	0.875	−1459.55	−124.75	
$c(5 \times 2)$ 8CO A	1.0	−1123.68	+193.29	•
$c(6 \times 2)$ 9CO A	0.9	−1456.35	−125.44	•
additional $c(4 \times 3)$ structures:				
$c(4 \times 3)$ 2CO	0.1$\bar{6}$	−2060.41	−123.91	•
$c(4 \times 3)$ 3CO A	0.25	−2057.90	−185.23	•
$c(4 \times 3)$ 3CO B	0.25	−1953.33	−159.09	•
$c(4 \times 3)$ 4CO A	0.$\bar{3}$	−1953.74	−212.25	•
$c(4 \times 3)$ 4CO B	0.$\bar{3}$	−2039.00	−240.68	•
$c(4 \times 3)$ 4CO C	0.$\bar{3}$	−1965.45	−216.16	•
$c(4 \times 3)$ 4CO D	0.$\bar{3}$	−2036.78	−239.94	•
$c(4 \times 3)$ 5CO C	0.41$\bar{6}$	−2038.86	−300.79	•
$c(4 \times 3)$ 5CO D	0.41$\bar{6}$	−1882.69	−235.72	•
$c(4 \times 3)$ 5CO E	0.41$\bar{6}$	−2029.14	−296.73	•
$c(4 \times 3)$ 5CO F	0.41$\bar{6}$	−2031.62	−297.77	•

Table 8.1: Complete pool of DFT input structures for the parameterization of the lateral interaction parameters within the cluster expansion for the CO binding energies. A black dot indicates the structures that have been included into the final set of configurations for the parameterization. Not included structures were either distorted or did undergo changes of the adsorption site upon geometry relaxation.

Formation Energies of the Structures Included in the Set

As already introduced in subsection 5.3.1 the formation energy of a specific adsorbate configuration with respect to a separation into a fully covered monolayer phase and a clean surface is a useful measure for adsorption processes that gives insight into the stability of a certain configuration. For this reason, the calculated formation energy for all DFT data points (GGA-PBE) with respect to the $c(2 \times 2)\,2\,CO$ A and the clean Pd(100) surface is plotted

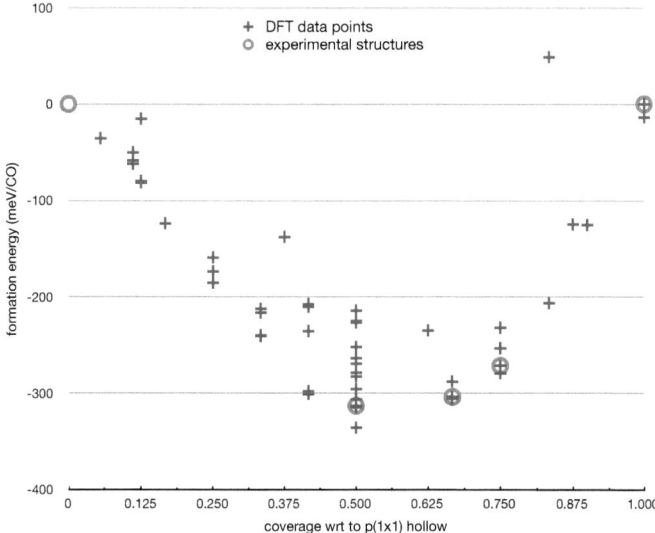

Figure 8.10: The calculated formation energies with respect to the $p(1 \times 1)\,2\,CO$ monolayer configuration and the clean Pd(100) surface of all adsorption configurations calculated at the GGA-PBE level. The green circles are representing the calculated formation energies for the experimentally characterized adsorption configurations.

Also included in this plot is the determined formation energy for the experimentally characterized adsorption configurations on the level of the GGA-PBE calculations. Interesting to note here is the fact that obviously some of the structures in the DFT database (blue crosses) appear to be more stable than the formation energies of the experimentally found adsorption configurations (green circles). For the definition of the *convex-hull*, however, always the configuration with the lowest formation energy was taken into account.

8.3.7 CV Scores for Different Sub-Sets of Interaction Parameters

For the determination of the optimal set of interaction parameters the selection was based partly on an interpretation of the underlying physical principles of the problem followed by a combinatorial based selection with the CV score value as the determining quantity. As a starting point a quite large set of different sorts of interaction parameters was set up, see above Fig. 8.7. Then starting from a first subset with only pair and trio interactions, over a second subset including also quattro interactions, a third one including also quattro interactions and quinto interactions to finally a parameterization with the full interaction set was performed and the corresponding cross validation score was calculated. In addition to this score value, also the former directly parameterized values for the interaction parameters V_1^p and V_2^p were taken as an indication of the quality of the resulting best fit for the interaction parameters. The different fitting processes were always performed using the selected set of 37 DFT input structures marked in the overview table 8.1 above. Including also distorted structures led to significantly (one order of magnitude) higher CV score values.

It can be seen from the results presented in table 8.2 that a pure expansion into pair and trio interaction parameters is not sufficient for the set of DFT input structures used in this work. The CV score value as a measure of the average error in the binding energy for a structure predicted by the evaluation of the cluster expansion lies in the order of 95 meV, which is clearly not acceptable. In addition, also the assigned energy values to the «benchmark» interaction parameters V_1^p and V_2^p are clearly deviating from the direct determined values. This behavior undergoes a major improvement when also four-particle many body interactions of the parameter set are included. The CV score value drops to 37 meV and to both pair interaction parameters their directly determined energy values are assigned within the parameterization. This assignment seems to be stable from this point on against the further inclusion of more higher order interaction figures.

	Interaction Set composition					
Identifier all energy values in meV/CO	Pairs / Trios	Pairs / Trios / Quattros	Pairs / Trios / Quattros / Quintos	Full Set	Optimized Set 1	Set 2
CV Score (in meV)	95.60	36.92	31.81	27.72	24.54	24.52
V_1^p	+211.1	+188.6	+188.6	+188.6	+188.6	+188.6
V_2^p	+51.7	+113.2	+113.2	+113.2	+113.2	+113.2
V_3^p	-27.9	-69.4	-75.2	-74.1	-37.1	-36.5
V_4^p	-59.5	-56.1	-45.5	-18.9	-31.3	-32.0
Continued on next page ...						

Identifier all energy values in meV/CO	Pairs / Trios	Pairs / Trios / Quattros	Pairs / Trios Quattros / Quintos	Full Set	Optimized Set 1	Set 2
V_5^p	-22.8	+1.0	+0.1	-3.4		
V_6^p	+13.0	+27.9	+17.1	+18.7	+28.9	+29.8
V_7^p	-18.5	+31.9	+14.3	+24.3	+23.6	+23.3
V_{11}^p	-6.8	+0.3	+0.4	+0.4		-1.8
V_1^{tr}	+380.8	+132.7	+137.7	+149.3	+135.6	+132.3
V_2^{tr}	+4.7	-144.2	-116.8	-131.1	-149.3	-148.4
V_3^{tr}	+69.8	+70.9	+39.4	-14.1	+10.6	+11.9
V_4^{tr}	+211.9	+30.2	-74.0	-72.9		
V_5^{tr}	+50.0	+16.5	+18.2	+17.8	+22.8	+21.7
V_6^{tr}	+10.7	+33.6	+39.8	+39.7		
V_7^{tr}	+30.6	-15.8	+20.9	+5.7	+11.4	+12.7
V_1^{qua}		+88.5	+91.8	+99.6	+90.4	+88.22
V_2^{qua}		+176.9	+183.6	+199.1	+180.8	+176.5
V_3^{qua}		+68.7	+55.0	+72.8	+63.4	+63.4
V_4^{qua}		+32.0	+36.4	+39.0		
V_5^{qua}		+65.9	+75.9	+46.8	+43.6	+43.4
V_6^{qua}		-8.0	-4.0	-20.8	-12.5	
V_7^{qua}		+79.8	+45.3	+57.2	+67.7	+66.9
V_8^{qua}		+132.5	+222.7	+202.0	+159.0	+163.4
V_9^{qua}		+6.6	-26.8	-12.9	-8.4	-10.3
V_{10}^{qua}		+87.3	+47.7	+62.7	+74.7	+73.5
V_{11}^{qua}		-2.1	-16.6	-12.3	-15.3	-16.4
V_{12}^{qua}		-42.6	-16.5	-27.0	-33.6	-32.6
V_{13}^{qua}		-2.7	-4.1	-4.1	+4.7	+4.9

Continued on next page ...

8.3. The Lateral Interactions of Adsorbed CO on Pd

	...continued from previous page					
	Interaction Set composition					
Identifier all energy values in meV/CO	Pairs / Trios	Pairs / Trios / Quattros	Pairs / Trios Quattros / Quintos	Full Set	*Optimized* Set 1	Set 2
V_1^{quin}			+14.2	+17.9	+21.1	+20.9
V_2^{quin}			+81.8	+89.8	+55.2	+55.6
V_1^{six}				+2.2	-16.3	-9.8
V_2^{six}				0		
V_3^{six}				+2.2		-9.8
V_1^{septem}				+20.4	+20.1	+19.5

Table 8.2: This table shows the different energy values assigned to the various interaction parameters (in meV/CO) of the pool of interaction parameters as resulting during the cluster expansion parameterizations. At the top of the table, the corresponding values for the cross-validation scores (in meV) are given.

Using the full set of interaction parameters results in a reasonably low cross validation score of 28 meV. An important aspect to mention here is the fact that upon further extension from the subset with pairs, trios and quattros to the full set, the assigned energy values of the individual interaction figures present a remarkably consistency. Not only that in almost all cases the ordering in terms of repulsive (positive) and attractive (negative) interaction parameters stays constant, but also the specific energy values change only little.

A further extension of the full set of interaction parameters with more longer range pair and trio parameters added to the set did not lower the cross validation score values, but led to an increase instead. Although not strictly systematically, this increase was taken as an indication that the full set represents a reasonably good and flexible enough set of interaction parameters. The two sets labeled *optimized* in the overview in table 8.2 however, were found using a combinatorial approach. Every possible combination of $33, 32, 31, 30, 29, 28, 27$ interaction parameters selected from the full set was generated and the corresponding CV score calculated. The two sets presented here are the two ones with the lowest CV scores with 28 and 27 parameters. Also for these the assigned energy values show a remarkable consistency with the values of the other expansions.

An exhaustive combinatorial generation of all possible subsets was computationally not possible. Generating all possible combinations $\binom{n}{k}$ from $k = 1 \ldots n$ would require the evaluation of $\sum_{k=1}^{n} \binom{n}{k} = 2^n$ cross validation scores. This turned out to be impossible for the size of 34 of the interaction parameter set. A systematic generation of combinations with a lower number of interactions (up to 24) based on the optimized set with 28 interaction parameters did also not lead to improved cross validation scores.

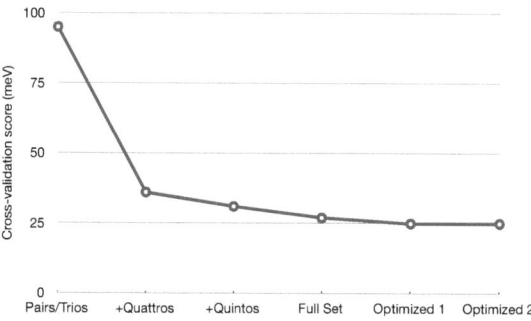

Figure 8.11: Shown here is the development of the calculated cross-validation score for the different sets of interaction parameters included in the cluster expansion. The effect of over-fitting can be nicely seen in the further decay of the CV score value for the optimized sets 1 and 2.

8.3.8 Ground-State Validation

The cluster expansion should not predict configurations for which the binding energy is lower than the DFT energy value for that particular coverage. This requirement should ensure that all ground state energies included in the parameterization have been calculated with the higher level method. However, in the case study presented here, this requirement was difficult to fulfill with the set of 37 selected DFT input structures, due to the observed strong relaxation effects for high density structures.

We checked for the ground state of the cluster expansion with an approach in which for two different unit cell shapes $c(4 \times 3)$ and $c(6 \times 2)$ all possible adsorbate configurations were generated and the corresponding binding energy predicted by the cluster expansion was calculated. Larger unit cell sizes were again beyond the limit of computational feasible tasks.

As visible in Fig. 8.12 and Fig. 8.13 below a coverage of $\theta = 0.5$ ML the expansion into the set of 28 interaction parameters with the lowest CV score in table 8.2 above performs quite well. None of the predicted binding energies falls below the accuracy limit of $(E_{\text{bind}}^{\text{DFT}} - 25 \text{ meV})$

8.3. The Lateral Interactions of Adsorbed CO on Pd

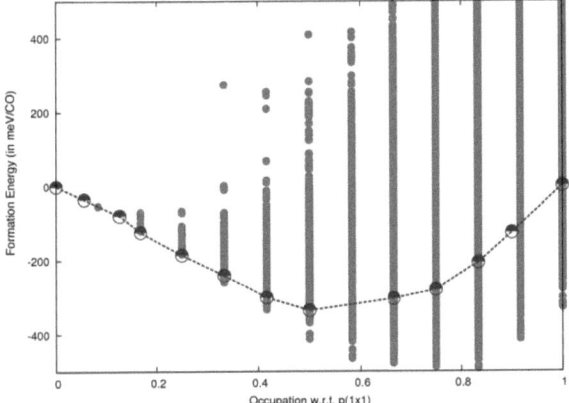

Figure 8.12: Ground state validation for the cluster expansion in the $c(4\ 3)$ unit cell. Visible in this plot is that for coverages below 0.5 ML all binding energies predicted by the cluster expansion are within the range of energy for a certain coverage given by the DFT value plus minus the uncertainty of approximately 30 meV of the cluster expansion as given by the cross validation score. For coverages higher than half a monolayer, the expansion into the optimal set of interaction parameters breaks down.

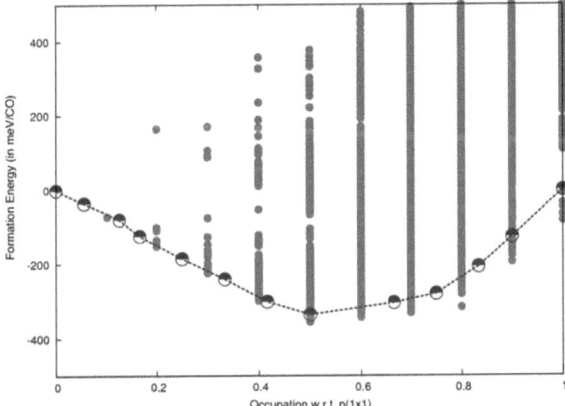

Figure 8.13: Ground state validation for the cluster expansion in the $c(6\ 2)$ unit cell. It is visible, that for this unit cell shape the cluster expansion performs better in the high coverage regime.

and therefore into the accuracy range predicted by the cross validation score. However, for coverages above the half monolayer, combinations with dramatically lower binding energies are predicted.

This observation is interpreted as a breakdown of the cluster expansion in the high density range. The difference between the predicted binding energies and the binding energies of configurations that have been observed experimentally in the higher coverage regime is too large to be an indication of a wrong selection of DFT ground structures. Clearly the parameterization for higher coverages is not sufficient with the selected set of 37 DFT structures. However, as was observed in the performed geometry relaxations for more densely packed structures, the influence of relaxations effects increased tremendously for these configurations. As already mentioned, such relaxations cannot be covered by a cluster expansion approach with statically defined interaction figures. Moreover, in cases were even adsorption site changes were observed due to the strong repulsive interactions the basic assumption of site-specific adsorption breaks down.

In addition, checking for two different unit cell shapes revealed that for the $c(6 \times 2)$ unit cell, the cluster expansion performs apparently better also in the high coverage regime. It can be seen that for this unit cell shape also for high coverages the deviation from the convex hull are less dramatic. This is an interesting observation. Recalling the unit cell shapes of the experimentally characterized adsorption structures shows that the unit cells of the denser ordered structures are realized by an elongation of the unit cell dimension in x direction, whereas the y size remains unchanged. The observation of the smaller deviations for a unit cell shape closer to the experimental structures is therefore a positive result with respect to the predictive power of the cluster expansion into the 28 parameters presented here. On the other hand this observation allows a useful methodological insight into the ground state validation procedure taken here and also in other works (e.g. in [18] only a square (4×4) unit cell was taken as reference for the ground state validation). For the case of our system it seems to be important not to concentrate only on one unit cell shape for the validation. This insight can be a useful information for further studies employing the method of cluster expansion for other systems.

8.4 Conclusions and Limitations

In summary, a cluster expansion for the adsorption of CO on bridge adsorption sites into a set of 28 lateral interaction parameters has been developed. This expansion bears an average predicting error of about 25 meV for the binding energies below a coverage of $\theta = 0.5$ ML. In comparison with recent publications this value is significantly higher than the one found by Zhang et al.[99] for the oxygen adsorption on the same Pd(100) surface, but still compares well to other published CV scores[88].

Above 0.5 ML this cluster expansion breaks down and does not predict useful binding energies anymore. This breakdown is assigned to the strong relaxation effects of this particular adsorbate. The strongly repulsive interactions between the adsorbed CO molecules lead to

8.4. Conclusions and Limitations

bending relaxations of the adsorbate axis with respect to the surface normal. This effect was observed in the performed geometry relaxations on the GGA-PBE level and its impact on the resulting binding energies can just be included as an uncertainty in the presented cluster expansion, since no parameter has been introduced to account for this particular energy contribution to the binding energies. In addition, locally dense structures — as necessary for the parameterization of short range many body interactions, turned out to be unstable against the switching of the adsorption site in the geometry relaxation. This behavior obviously is also not included in a cluster expansion focused on bridge site adsorption. Therefore, we can conclude that the behavior of the system in the high coverage regime will most likely be determined by the very strong repulsive interaction which may even lead to a population of less favorable adsorption sites. Our extracted first nearest neighbor interaction and second neighbor interaction parameters and also the attractive third one agrees well with parameter deduced from experiments for another surface facet[150] and also with previously reported values[161].

We were therefore able to gain useful insight into the properties of this particular adsorption system in the low coverage regime with our approach. We could show that a simple expansion into only pair and trio interactions is apparently not sufficient and that higher order many body interactions are essential for this system. In addition to this useful insight, we followed an approach that for the first time derived the whole set if interaction parameters from first-principles starting from a microscopic perspective instead of fitting to experimental data. Therefore, our extracted interaction parameters are strictly connected to interactions between adsorbates on a molecular level. On this level, the majority of interaction parameters is clearly repulsive, which agrees well with the expectations for this system and all former investigations.

In addition, we could show that the unit cell shape can play a role in the ground state validation procedure usually employed. The observed smaller deviations in the predicted binding energies for higher coverages in a unit cell which shape is closer to the experimentally found one shows that the unit cell shape has an important influence for this procedure and a cluster expansion parameterization in general.

Chapter 9
Submonolayer Heteroepitaxial Growth of Cu Islands on Ag(100)

nel mezzo del cammin di nostra vita
mi ritrovai per una selva oscura
che la ritta via era smarrita

Dante Alighieri

FAST scanning tunneling microscopy reveals an unusual structure and mobility of smaller Cu-islands on clean Ag(100) at room temperature.[19] Whereas islands containing more than 80 atoms exhibit a diffusion and decay behavior similar to the one of homoepitaxed Cu and Ag islands on Cu(100) and Ag(100), respectively, smaller islands show a more complex structure with Cu atoms seemingly adsorbed in bridge sites, and a diffusivity and decay time that is significantly higher than any previously measured value. Motivated by these experimental findings and by the fact that a metal on metal system represents an nice showcase for a different interplay regime between lateral interactions in the adsorbate layer and substrate–adsorbate interactions, the second chapter in the results part of this thesis will deal will the heteroepitaxial growth of small amounts of Cu on the Ag(100) surface.

The Cu island structures and the physical driving forces responsible for certain reconstructions will be studied and analyzed within a multiscale approach, particularly covering several length scales. First-principles based calculations will be used as a starting point for an analysis involving structural relaxations and extracting electronic information by means of STM simulations in order to compare to the experimental data.

Based on the extracted information about the substrate-adsorbate interaction and the adsorbate–adsorbate pair interaction we extend the simple Frenkel-Kontorova model to three dimensions and investigate the configurational space of island structures with a simulated annealing approach. The insight gained by this approach then forms the motivation and justification to go beyond a purely pair interaction description of the adsorbate–adsorbate interactions with an embedded atom method (EAM) based potential description.

After the validation of the EAM potential against a set of density-functional calculations, we go beyond the microscopic scale and access very large system sizes. We converge finally to island size ranges which are of the order of the experimentally investigated system sizes. For those systems we are able to derive an explanation of the physical driving force behind the reconstruction observed in the experiments based on an analysis of our simulations.

9.1 Lateral Interaction in Metal Heteroepitaxy

In the previous example (chapter 8) the interaction situation that we were facing was the one, where the bonding to the substrate was large compared to the lateral interactions between the adsorbates. Speaking in terms of the denominators introduced in the first chapter of this thesis, this was the case $E_b \cdot a_s \gg k/a_c$.

The other extreme case, namely $E_b \cdot a_s \ll k/a_c$, describes the rather special situation of a *floating adlayer*, as one would find for example for the case of physisorbed rare gas adlayers on cold metal surfaces.[164] The theoretical treatment based on current density-functional theory is complicated due to the importance of van der Waals (vdW) type interactions as the governing adsorbate–adsorbate interactions in those systems. Most currently used density-functional implementations lack a proper treatment of these type of particle interactions. However, emerging new schemes also allow for a more and more accurate treatment of those as well and have proven to be quite successful in covering the fundamental physics[165,166].

A much more common situation is described by $E_b \cdot a_s \leq k/a_c$, which means that the interactions of the adsorbates amongst each other are almost of the same strength as the bonding to the surface, thus leading to a strong competition between those two interactions. This situation is usually referred to by the term *epitaxy*, which describes a situation, where the structural integrity of the adsorbate layer material is of an almost equal energetic importance for the complete system as the adsorbate-substrate interaction.[1] Obviously, this leads to configurations where the adsorbate atoms are relatively closely arranged to each other and the lateral interactions between those atoms determine an adsorbate «lattice» constant. Due to the already mentioned fact that this lattice constant will in almost all heteroepitaxial cases differ from the substrate lattice constant, typically one would expect for this situation the formation of some more or less strongly reconstructed island structures on the surface. In addition to it being more often found in adsorbate systems, this situation is also amenable to the descriptions by common density-functionals, since the physical origins of the governing interactions between the adsorbates are properly described for those cases.

By means of experimental techniques such as the slow evaporation of a metal in ultrahigh vacuum it is possible to adsorb low concentrations of Cu atoms on Ag(100) surfaces. This has been investigated by fast scanning-tunneling-microsopy (STM) techniques by Christopher Zaum[19]. The experiments showed the formation of island structures that exhibit rather unusual reconstruction features. Therefore, this system will be the showcase for the second application part of this study, representing a case of a strong competition between lateral interactions and adsorbate-substrate interactions.

9.2 The Copper-Silver Alloy System

The phase diagram of the binary system copper/silver shows a wide miscibility gap and has a clear tendency for segregation.[167,168] No stable solid solution phase of this system is known or has been experimentally synthesized so far. Measurements of the formation enthalpies

9.2. The Copper-Silver Alloy System

of the metastable solid alloy phases prepared by ball mining experiments showed positive values over the full range of Cu concentrations.[169]

Despite the miscibility gap in the bulk, it has been shown that the growth of Cu adlayers on a silver surface is possible. In those experiments initially island formation is observed either for the case of electrodeposition or vacuum deposition of Cu on the surface. This initial stage is then followed for higher surface concentrations of copper by adlayer growth in form of a strained pseudomorphic bcc phase up to nine to ten monolayers. Further deposition goes along with a structural transformation of the newly build layers caused by a reduction of the measured Cu–Cu bond distance, which leads to the formation of the more stable fcc structure of copper from this layer on and to the formation of large Cu crystallites on the surface.[170,171] In addition, the formation of a surface alloy thin film has been observed experimentally when the binary mixture of copper and silver is deposited on a different substrate.[172]

The observed initial island formation can be interpreted as a manifestation of a segregation tendency of this particular system similar to what is observed for the solid phase solutions. In addition, the island formation is clearly an indication for rather strong lateral Cu–Cu interactions compared to the Cu–Ag adsorbate-substrate interactions. Interest for this system can be found for its peculiar catalytic properties.[173] Lateral interaction between the adsorbed particles have been found to influence the island formation and the dynamics of adsorbate diffusion. This has been attributed to a combination of long- and short-range effects.[174,175] These mutual interactions of island atoms have also been used to explain the mentioned fact, that surface alloy formation is possible for thin films of this binary system on top of surfaces of other metals as substrate.[172,176]

From a structural point of view, Cu and Ag exhibit a rather strong mismatch in the equilibrium bulk lattice constants $a_0^{exp}(Ag) = 4.0853$ Å vs. $a_0^{exp}(Cu) = 3.6149$ Å for the experimental values taken from Ref. [177]. Density-functional theory reproduces this lattice mismatch ratio quite nicely $a_0^{GGA-PBE}(Ag) = 4.1377$ Å vs. $a_0^{GGA-PBE}(Cu) = 3.6258$ Å, see also A.1.1. These differences in the bulk lattice constants are naturally also resembled in the corresponding surface lattice constants, which in turn leads either to strained structures in the case of pseudomorphic growth or to structural rearrangements in the case of heteromorphic growth behavior.

9.2.1 Experimental Observations of the Island Structures

In the experiments that present the basis for the theoretical investigations of this thesis the following results have been observed in a series of Fast Scanning Tunneling Microscopy (STM). Fig. 9.1 shows an overview STM scan of the surface and a linescan height profile generated by following the green indicated line in the overview picture. After the preparation and cleaning of the Ag(100) surface in the experiments the Cu atoms are evaporated at an evaporation rate between 0.001 ML/min and 0.01 ML/min at room temperature ($T = 293$ K).

96   Chapter 9. Small Cu Islands on Ag(100)

Several sub-monolayer coverage situations with coverages between 0.003 ML and 0.13 ML are prepared in this way. The majority content of the evaporated copper is aggregating in island structures, a minor fraction adsorbs as fast diffusing single atoms. The shape of these islands is almost quadratic in the STM images, as can be seen in Fig. 9.2, for example. Also visible from the so-called differential plot shown on the right side of Fig. 9.2 are the

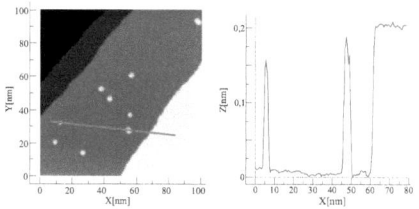

Figure 9.1: Experimental STM image (left) and linescan following the green line for adsorbed Cu structures on the Ag(100) surface. Taken from Ref. [19], the experimental parameters were: $I_{tunnel} \approx 0.8\,nA$ for a voltage of $U_{tunnel} \approx 800\,mV$.

positions of the Cu atoms, some of which appear to occupy the *bridge* adsorption sites of the underlying Ag(100) substrate lattice. Two height regimes are observed for islands of different si‍z   lower
appar‍e   STM images 3d - Cu to bridge assigned   cantly
higher

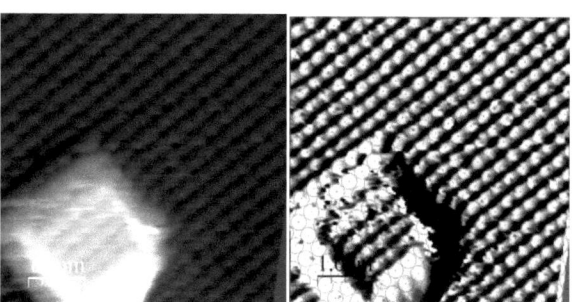

Figure 9.2: Nearly atom-resolved STM image of a typical quadratic Cu island structure for the sub-monolayer growth experiments of Cu islands on the Ag(100) surface. The picture is taken from Ref. [19], the experimental conditions have been: $I_{tunnel} \approx 0.8\,nA$ for a voltage of $U_{tunnel} \approx 800\,mV$ at a temperature of $T \approx 220\,K$. On the left side the actual STM image is shown. The right side shows a *Differential Plot* of the STM signal, in which the assigned positions of the adsorbed Cu atoms are marked as bright yellow spots.

A series of islands showing different island sizes have been characterized in the experiments and the results of this are plotted in Fig. 9.3. Apparently two growth regimes exist, which depend on the island size. Smaller island structures below an area of about

9.2. The Copper-Silver Alloy System

$A \lesssim 7\,\text{nm}^2$ seem to occupy the bridge adsorption sites of the underlying Ag(100) substrate lattice, whereas the islands with areas $A \gtrsim 7\,\text{nm}^2$ appear to prefer Cu atoms sitting in the fourfold hollow sites. Important to note here is, however, that only some of the island structures have actually been investigated in atomic resolution STM measurements and the assignment of the adsorption site has been mainly carried out by the assumed correlation between the average apparent island height as an indication for the preferred adsorption site of all the island atoms. Whether this strict correlation holds for all island atoms in all cases remains a matter of debate and interpretation. In this situation, our study addresses

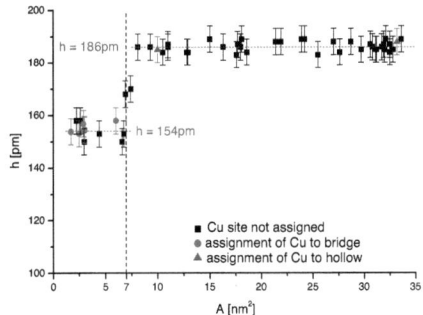

Figure 9.3: Shown in this summary graph (taken from Ref. [19]) are the measured average apparent island height against the measured island area. Clearly, two regimes of different island heights are visible. For some of these islands, the preferential adsorption sites have been determined in atom-resolved STM images. Smaller island structures below an area of about $A \lesssim 7\,\text{nm}^2$ seem to occupy the bridge adsorption sites of the underlying Ag(100) substrate lattice, whereas the islands with areas $A \gtrsim 7\,\text{nm}^2$ appear to sit in the fourfold hollow sites.

this question from a theoretical point of view using predictive-quality simulations for which the assignment of certain atoms to certain lattice positions is perfectly unambiguous.

The major questions that we try to answer in these investigations are:

- What is the physical driving force behind the apparent reconstruction tendency?
- Why is there a structural difference and what kind of difference is there between small and large islands?
- Can the critical island size be rationalized from a theoretical point of view?

These questions will be addressed within the framework of a multiscale modeling approach. The size-dependent reconstruction behavior of square monolayer Cu islands on a Ag(100) surface will be studied by density-functional theory and two different coarse grained model descriptions. A driving force of the reconstruction behavior will be presented and our results will be compared to the experimental findings. The parameters for the first-principles based Frenkel-Kontorova model will be explained and it will be shown how they have been derived from the electronic structure energetics. The question of the global minimum island structure will be addressed by an extended structure search. STM simulations will allow direct comparison of some island features of the theoretical structures with the experimental results.

9.3 Theoretical Investigation

9.3.1 Density-Functional Theory as a First-Principles Starting Point

Stable Island Structures

We started our theoretical investigation of this system by optimizing the geometries and calculating the adsorption energies of a monolayer of Cu placed at a Ag(100) hollow position in contrast to a monolayer of Cu atoms placed on (100) bridge positions. These calculations effectively mimic the bonding properties of adsorbed Cu atoms in very (infinitely) large Cu islands. In full agreement to previous experimental[171] and theoretical investigations adsorption into the fourfold coordinated hollow-site turns out to be significantly more stable than the bridge site, namely −0.434 eV/Cu atom for the periodic (1 × 1) calculations. For very large islands, there is thus no doubt that the stable position of adsorbed Cu atoms is the high-symmetry hollow site.

As second step we focused on a square (4 × 4) Cu island containing 16 Cu atoms. Placing this island into a (6 × 6) surface unit-cell implies that the island is separated from its periodic images by 4 atomic rows of the Ag(100) surface in every direction. From test calculations of the same island setup in a (7 × 7) surface unit-cell, with a correspondingly enlarged separation between the island periodic images, we conclude from the unsignificantly altered binding energy per Cu atom that the Cu islands are sufficiently decoupled laterally from each other already in the setup with the smaller surface unit-cell.

We now define the stabilization energy per Cu atom for each island structure as:

$$\Delta E/\text{Cu} = \frac{E_0^{\text{DFT}}(\text{all-hollow structure}) - E_0^{\text{DFT}}(\text{bridge/reconstructed structure})}{\text{nr of Cu atoms}} \quad (9.1)$$

Following this definition, a positive ΔE values describes a situation in which the reconstructed island structure is more stable, whereas a negative sign points to a situation in which the configuration with all Cu atoms adsorbed in hollow sites is the more stable one.

Placing the island Cu atoms into hollow positions, we obtain only insignificant atomic relaxations, i.e. all Cu atoms prevail essentially in the fourfold coordinated hollow sites. If we place the Cu atoms into bridge sites and constrain them laterally to these positions (i.e. allow only vertical relaxation), the energetics of the preceding (1 × 1) surface unit-cell calculations is almost perfectly reproduced: Again, the island with all Cu atoms in hollow sites is more stable, with a higher average binding energy per Cu atom of −0.476 eV/Cu atom (cf. to −0.434 eV/Cu atom in the infinite island size limit).

9.3. Theoretical Investigation

Figure 9.4: Top and side view of the reconstruction 4 × 4 Cu island, with some Cu atoms residing close to bridge positions due to the formation of hexagonal structural motifs within the Cu island.

However, when we allow the Cu adlayer atoms of the bridge configuration to relax unconstrained into x, y and z directions, we observe strong relaxations and a reconstruction of the entire island structure. The resulting island geometry is shown in Figure 9.4 and features several Cu atoms in positions close to the nominal bridge sites of the underlying Ag(100) surface. As furthermore apparent from the side view, these atoms protrude now significantly from the overall island height. Intriguingly, the structural relaxation is accompanied by a significant energy gain, leading to a higher average energy per Cu atom in the reconstructed island compared to the island structure with all Cu atoms in hollow sites, namely +0.083 eV/Cu atom. In other words, at least in the direct comparison of the two square (4 × 4) island structures the reconstructed structure is now more stable than the one where all Cu atoms remain in the local minimum configuration presented by the occupation of purely hollow adsorption sites of the Ag(100) surface. Comparing the total energy results obtained for the (1 × 1) Cu monolayer and the square (4 × 4) island we can therefore conclude on a transition of the most favorable island structure with island size, with small islands preferring a strongly reconstructed and buckled geometry with some Cu atoms close to bridge sites and large islands preferring a flat geometry with Cu in the fourfold hollows.

From this starting point, we studied a series of bigger square island structures (5 × 5, 6 × 6, 7 × 7 – always appropriately enlarging the surface unit cell to keep a lateral separation of the islands of at least 4 lattice spacings of the Ag surface).

As before, for each size we performed a full geometry optimization for a starting structure with all Cu atoms placed at the Ag(100) hollow sites and another relaxation for a starting configuration with all Cu atoms initially placed in bridge sites. We observe in all studied hollow structures only minor relaxations of the Cu atoms, mostly adjusting the initial adsorption height to the minimum energy one. In contrast, all studied bridge structures relax into a strongly reconstructed state, similar to the one described already for the (4 × 4) structure. The resulting stabilization energies per Cu atom in the island structure with respect to the all-hollow configurations are listed in Table 9.1. Al-

Cu island size	ΔE/Cu in eV
3 × 3	−0.002
4 × 4	+0.083
5 × 5	+0.045
6 × 6	+0.070
7 × 7	+0.065
monolayer	−0.434

Table 9.1: Stabilization energy per Cu atom for initially square Cu islands on Ag(100) after a DFT-BFGS geometry relaxation.

though at first glance these numbers seem small, one has to keep in mind that these values are always normalized for the number of island atoms. Therefore, even the small energy gain of +0.045 eV for the (5 × 5) island structure amounts to a total energy difference at the level of the employed GGA-PBE functional of +0.045 eV/Cu · 25 Cu = 1.125 eV = ΔE_{total} for the two structures. This represents a significant stabilization energy, even when taking the uncertainties accompanied with the assumptions underlying the approximate GGA-PBE functional into account.

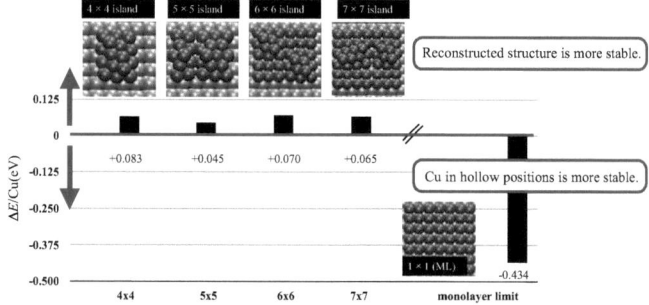

Figure 9.5: Most stable island structures after geometry relaxation for different island sizes with their ΔE values indicating the most stable configuration. Also shown is the energy difference for the monolayer configuration, which clearly favors the occupation of the fourfold hollow site of the Ag(100) surface. Therefore, from this energetics one can conclude that for larger and larger island structures (beyond the scope of current day DFT implementations) a turnover in the stability properties must occur.

In Figure 9.5 these results are represented in a graphical way, also showing the geometrical shapes of the resulting reconstructed island structures (enlarged pictures of the relaxed island structures can be found in Appendix C). A closer look at these also explains the rather «low» stabilization energy for the (5 × 5) structure. This island size reconstructs to a rather open structure, and is not yet big enough to allow the formation of the two (111)-facet-like structures visible for the bigger structures. However, it is already too large to form only one of those facets as it is the case for the (4 × 4) structure. This naturally results in a situation where a part of the island structure is forced to remain in a rather unreconstructed structure or in a rather unfavorable half-reconstructed state, therefore resulting in a smaller overall energy gain by the reconstruction.

9.3. Theoretical Investigation

The Driving Force for the Reconstruction

In an analysis of the physical driving force behind this reconstruction behaviour, we start by recalling the large lattice mismatch of the fcc bulk lattice constant of Ag and Cu of about 13% in the (100) direction, see the values in Table 9.2.

		$a_0(Cu)/a_0(Ag)$ in %
$a_0^{GGA}(Ag) = 4.138$ Å	$a_0^{GGA}(Cu) = 3.626$ Å	87.6
$a_0^{exp.}(Ag) = 4.085$ Å	$a_0^{exp.}(Cu) = 3.615$ Å	88.4
$a_{100}^{GGA}(Ag) = 2.926$ Å	$a_{100}^{GGA}(Cu) = 2.564$ Å	

Table 9.2: DFT and experimental values of the lattice constants of Ag/Cu

If a Cu island would form step by step on a silver surface through the accumulation of more and more Cu atoms, two construction principles are possible. Either the single Cu atoms would sit all in hollow positions and adapt to the larger silver lattice constant, or the Cu layer would try to keep the optimal Cu–Cu distance and buckle consequently. We illustrate this concept in Fig. 9.6.

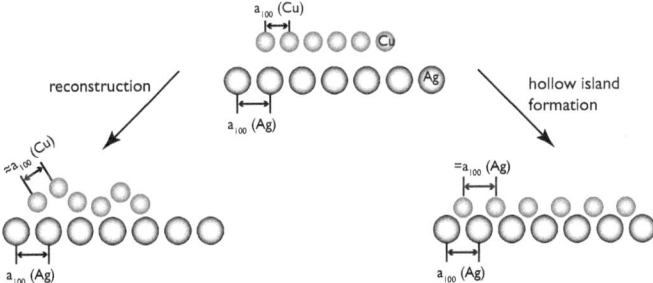

Figure 9.6: The large lattice mismatch between Cu and Ag as the physical driving force behind the reconstruction. The most stable island structure is determined by a competition between the gain in bonding energy to the substrate by occupying the most stable adsorption site and a contribution gained from an optimized Cu–Cu bondlegth within the island. The right pathway shows an island with maximum bonding energy gain, all Cu atoms are sitting in the hollow adsorption site. The left pathway leads to a buckled island structure, where some Cu atoms are pushed to unfavourable adsorption sites while optimizing the Cu–Cu distances within the island structure.

The driving force for the reconstruction is clearly a competition between optimized Cu–Cu and Cu–Ag bonds. Naturally, this is connected to either gaining or losing binding energy either from Cu–Cu bonds or Cu–Ag bonds. In the reconstructed $(n \times n)$ islands $(n = 4 \ldots 7)$ the Cu–Cu distances are significantly shortened and in fact approach the value corresponding to the Cu bulk lattice constant, a_0, quite closely, cf. Table 9.3.

	(4×4)	(5×5)	(6×6)	(7×7)
all-hollow islands	3.92 Å	3.96 Å	4.03 Å	4.03 Å
reconstructed islands	3.60 Å	3.57 Å	3.67 Å	3.69 Å

Table 9.3: Average island Cu–Cu bondlengths of the relaxed DFT structures.

As this is incompatible with the underlying Ag lattice structure, the Cu atoms are forced out of the fourfold-coordinated hollow sites corresponding to optimized Cu–Ag bonds. This incompatibility becomes more and more severe with increasing island size, i.e. larger islands would build-up an increasing amount of strain if attempting to preserve the short Cu–Cu distances. The size increase would go along with a higher and higher amount of energy contributions that could be gained from an optimized substrate-adsorbate bond situation.

Obviously, this naturally leads to a transition in the preferred island geometry: In small islands, the stronger Cu–Cu (compared to Ag–Ag) bonds dominate and the island reconstructs. In a rather intuitive conclusion, all reconstructed island structures reveal structural motives corresponding to hexagonal arrangements of Cu atoms resembling a close-packed (111) fcc layer, i.e. the final island geometry can be seen as the system's intention to form a compact Cu overlayer in the most favorable surface termination with Cu–Cu bond distances of this hexagonal structure on a square Ag(100) substrate with Ag–Ag distances. In larger islands this can no longer be supported due to the increasing strain resulting from the lattice mismatch. Eventually, the island therefore adopts a geometry characterized by all Cu atoms simply residing in the most stable adsorption site offered by the Ag(100) surface.

Also an island that is too small, cf. (3×3) energetics in Table 9.1, does not allow for a full reconstruction and stabilization of the island by a gain in Cu–Cu bonding energy. An island of this size is obviously not large and flexible enough to gain enough inner island bonding energy by the reconstruction to overcome the loss in bonding energy to the substrate. However, when starting from bridge positions initially also the (3×3) structure got significantly distorted after the geometry relaxation. Nevertheless, the energy gain by the reconstruction equals more or less the energy loss in binding energy.

9.3.2 Comparison to Experiment

STM Simulations of Island structures

In order to compare to the STM experiments, we performed STM Simulations of our calculated final structures within the Tersoff-Hamann[77] approximation, see section 3.4.

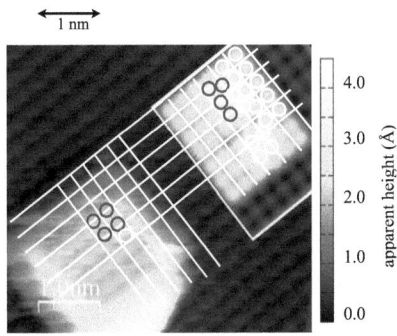

Figure 9.7: Comparison between simulated and measured STM images of an island of comparable size. Simulated STM is of a (7 7) reconstructed island structure.

We find that our simlulated STM images nicely reproduce the experimental features and also support the assignment, that some Cu atoms sit on bridge lattice positions of the underlying Ag(100) substrate. This is shown for the example of a (7 7) island in Fig. 9.7. For the STM simulation based on the island structure resulting from a DFT geometry optimization all Cu atom positions can be assigned uniquely, whereas the assignment of atom positions in the experimental STM image is based on a so-called Differential Plot of the STM measurements, as shown in Fig. 9.2. Only a few Cu atoms sitting very close to the nominal bridge sites of the underlying Ag(100) substrate lattice are marked by red circles for the experimental data in the picture. The STM simulation has been scaled such that the lattic structure of the surface equals the experimental conditions. The zero level of the apparent height coloring has also been adapted to the experimental one to allow comparisons of the colorcoded apparent height. Similar STM images have been calculated for all island structures resulting from the DFT geometry relaxations. In all cases bright spots of atoms sitting close to bridge positions could be identified. However, the biggest structure calculated, namely the (7 7) island structure, corresponded to the smallest experimental image in nearly atom-resolved resolution. For this reason, these two structures only have been compared here.

9.3.3 Trend Prediction by a Theoretical Separation of the Stabilization Energy

In an analysis of the energetics, we separated the total stabilization energy for the different island structure into a bonding contribution that arises from the bonding to the Ag(100) substrate and a bonding energy contribution arising from the inner island Cu-Cu bonding. For this comparison we define the bonding energy to the substrate as:

$$E_{bond}(\text{island}) \quad E_0^{DFT}(\text{island@surface}) \quad E_0^{DFT}(\text{island}) \quad E_0^{DFT}(\text{surface}) \quad . \quad (9.2)$$

We then calculate the difference in bonding energy between the reconstructed structures and the all-hollow configurations as

$$E_{\text{bond}} = E_{\text{bond}}(\text{reconstructed island}) - E_{\text{bond}}(\text{all-hollow island}) \quad . \quad (9.3)$$

In order to compare the pure island stability caused by the inner-island bonding, we further define the difference in inner island bonding as:

$$E_{\text{inner-island bond.}} = E_0^{\text{DFT}}(\text{reconstructed island}) - E_0^{\text{DFT}}(\text{all hollow island}) \quad . \quad (9.4)$$

Both contributions add up to the total stabilization energy given in Table 9.1. However, separating them provides insight into the interplay of these contributions and allows to identify already for the DFT database of island sizes a clear trend and the conclusion that at a critical island size a turnover is expected. This would happen, if the energy gain in stabilization energy by the bonding to the substrate overcame the energy gain of the Cu–Cu bonds within the island structures. This conclusion is illustrated in Fig. 9.8.

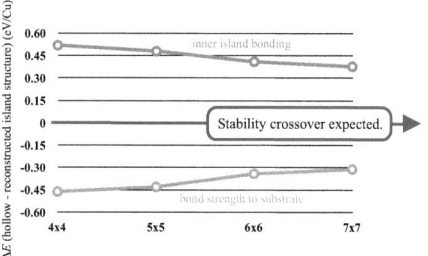

Figure 9.8: Separation of the island stabilization energy into a contribution arising from the bonding to the substrate and one arising from inner-island bonding.

Island Mobility

As a corollary, the analysis of the island stabilization in terms of bonding to the subtrate and inner-island bonding also provides a natural rationalization for the experimentally found enhanced mobility of the smaller reconstructed Cu islands compared to the larger unreconstructed ones, cf. Fig. 9.9. Shown is the diffusion constant D of the islands extracted from a movie of experimental STM images tracking the motion of a specific island and assuming a Brownian motion[178,179] of islands. The diffusion constant should generally be proportional to the area A of the islands and the temperature[180]:

$$D \propto \exp\left(-\frac{E}{k_{\text{B}} \cdot T}\right) \cdot A^{\alpha} \quad , \quad (9.5)$$

where E describes the activation barrier for the diffusion event, k_{B} is the Boltzmann constant, A the island area and the exponent α describes the dominant microsopic nature of the diffusion process.[180,181] One usually distinguishes (i) periphery/edge diffusion *(PD)* with $\alpha = 3/2$, (ii) terrace diffusion *(TD)* with $\alpha = 1$, and (iii) evaporation and condensation limited

9.3. Theoretical Investigation

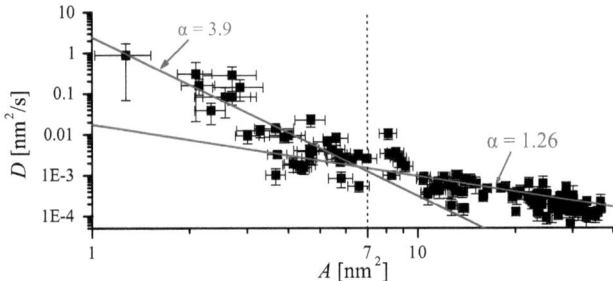

Figure 9.9: Island mobility as function of the island size as extracted by movies taken in the experiments reported in [19]. The given coefficients α clearly distinguishes two diffusion regimes. (For a discussion of the origins of α see text)

diffusion *(ED)* with $\alpha = 1/2$. It may be questionable from a mechanistic point of view based on an atomistic description, whether the diffusion of an island on a surface is in all situations due to mass transport in the form of single atom events. This assumption is the basis for the relation given in Eq. 9.5. However, also concerted mechanism have been identified for island diffusion which naturally are not covered by this proportionality.[182,183] In particular, recent theoretical studies identified a high probability of those kind of diffusion events for the special system of Cu island clusters on a Ag(100) surface.[184,185]

The investigation of a diffusion mechanism even with a computationally undemanding potential description is a challenging task, since the timescales one has to cover to detect a single diffusion event can easily extend tens of nanoseconds[183]. Nevertheless, the analysis of the stabilization energy contributions for the islands within the DFT energetics offers an appealing rationalization for a higher mobility of smaller islands in comparison to the larger island structures.

In the former, the main stabilization comes from strong intra-island Cu–Cu bonds with a concomitantly less important Cu–substrate interaction, whereas in the latter the main stabilization comes from the Cu–Ag interaction itself. This is e.g. reflected in the computed binding energies of the entire island to the Ag(100) substrate, cf. Table 9.4, which is with -0.978 eV/Cu atom smaller for the reconstructed (4×4) geometry than the -1.440 eV/Cu for the (4×4) island with all Cu atoms in hollow sites. This would suggest that the reconstructed islands are more prone to diffusion, simply because the average island-substrate interaction is smaller than in the case of the unreconstructed islands.

Moreover, one could argue that the probability of a collective diffusion process that would involve more than a single Cu island atom at once, is higher for a more compact island structure than for an island in which the island atoms are less bound amongst each other. This could be rationalized by the comparison of the inner-island bonding energies, which is higher for the reconstructed structures than for the all-hollow structures.

Cu island size	all-hollow E_{bond}/Cu	reconstructed E_{bond}/Cu
(3 × 3)	−1.479 eV	−1.103 eV
(4 × 4)	−1.440 eV	−0.978 eV
(5 × 5)	−1.390 eV	−0.959 eV
(6 × 6)	−1.440 eV	−1.100 eV
(7 × 7)	−1.423 eV	−1.112 eV

Table 9.4: Bonding energy of the island structure to the Ag(100) substrate. The reconstructed structures are always less strongly bound per island Cu atom.

9.3.4 Extending the Search Space of Island Structures

The results presented so far naturally rise the question, whether or not the reconstructions based on initially square and monolayer starting structures are the most stable island isomers for this system under the specific experimental conditions. As already stated in section 7.1, every energy minimization procedure — as used to relax the initial geometries of the DFT calculations — only relaxes a given starting structure to the next local potential energy surface (PES) minimum structure. This structure does not necessarily have to be the global minimum structure of the system. In fact, most likely it is not the global minimum. Another question that comes naturally to mind after a closer look at the trend prediction in island stabilization contributions as shown in figure 9.8 is at which size the turnover will be predicted by theory. Recalling from the experimental results, the critical size range was around $A = 7\text{nm}^2$ which would approximately correspond to a (9 × 9) or (10 × 10) square island.

However, even on modern supercomputers DFT calculations are a time consuming task. Moreover, one is limited in the size of the structures to calculate by the available computational ressources, like for example the requirements of main memory per node. In our case, the largest island structure that we were able to calculate is the (7 × 7) island in a (9 × 9) surface unit cell, leading to a total number of 292 transition metal atoms. This system was calculated on 512 POWER7™ CPUs, where the total time needed, to relax the starting geometries to a local minimum, was about 4 weeks. Therefore, calculating bigger structure are already unfeasable for pure time reasons. In addition, even the next bigger structure, i.e. a (8 × 8) island, extended already the available memory ressources of 2 GByte per supercomputer node.

These computational requirements clearly motivate the search for another model description that can be used for the energy calculation of arbitrary island structures. In addition,

9.3. Theoretical Investigation

the configurational phase space of the possible island structures should be more thoroughly explored by means of a global optimization procedure to scan for more stable structural conformers. Since the structure of the islands is evidently largely determined by the interplay between the mutual interactions among the island atoms and the adsorbate–substrate interaction, the model description to go beyond island sizes which are trackable by density-functional theory calculations must properly account for them in particular.

Recalling the qualitative model description of adsorbate–substrate and lateral adsorbate–adsorbate interactions within the Frenkel-Kontorova model from section 2.1.1, it comes clear that this model should be able to provide a starting point for a further investigation of larger island structures, which fulfills to a certain extend the two requirements. Total energies within the FK-model description are clearly easy to evaluate and the adsorbate–adsorbate interaction is explicitly accounted for by a specific pair potential interaction.

An Extended Frenkel-Kontorova Model Description

The original Frenkel-Kontorova Model[9,10] was designed as simplified model for the description of a one-dimensional chain of atoms interaction with a one-dimensional periodic substrate potential. Our extension of it to cover the full $3D$ space of reconstruction possibilities of island structures is basically straight foward and follows the line of some recent publications.[15,186]

In a first step, we parametrize a surface potential to describe the interaction between adsorbate atoms with the surface from first principles datapoints on the level of density-functional theory calculations. Secondly, the pair interaction potential for the adsorbate–adsorbate interaction is parametrized similarly. Finally, we perform global optimization attempts for different selected starting structures by a simulated annealing algorithm. The total energy of the system as given by the Frenkel-Kontorova model description represents the cost function to be minimized in those optimizations.

Surface Potential from First-Principles Obviously, for a $3D$ extension to the original Frenkel-Kontorova model the surface potential needs to be extended to $3D$ to properly describe the interaction with an adsorbate atom with the substrate at any point (x, y, z) above the surface. A very convenient way to gain a continuous description of a quantity that is only know exactly at a discrete set of grid points is to use spline interpolation procedures for all points between the known data points. This ensures that the input data points are reproduced exactly. Furthermore, the spline interpolation allows also to calculate certain order derivatives of the resulting interpolated spline function, avoiding discontinuities of the objective function.

In the Frenkel-Kontorova model, the adsorbate–substrate interaction of a single, isolated adsorbate particle with the surface potential is used to represent the binding contribution to the total energy as sum over all these individual binding energies. Therefore, for the interpolation's parameterization of the surface potential, this quantity must be evaluated for a single Cu atom above a Ag(100) surface for a grid of coordinates. Since we are interested in the isolated binding energy for the parameterization, we must ensure that in the setup of the first principles DFT calculations the periodic images of the unit cell are separated properly to avoid an influence on the binding energy caused by interacting periodic images.

As a consequence, the convergence of the quality of interest, the binding energy of a single Cu atom to two high symmetry adsorption sites (hollow and bridge) of the Ag(100) surface has been checked with respect to the lateral separation between periodic images provided by certain surface unit cell sizes. Comparing the energy values calculated in the $p(1 \times 1)$, the $p(2 \times 2)$ and the $p(3 \times 3)$ surface unit cell, it was found that the binding energy does not change significantly anymore between the $p(2 \times 2)$ and the $p(3 \times 3)$ cell. All subsequent data points have thus been calculated for a single Cu atom placed within a $p(2 \times 2)$ surface unit cell. In this surface unit cell one $p(1 \times 1)$ quarter of the surface area was sampled in its symmetry reduced triangle, as depicted in the left side of figure 9.10. The range between $x \in [0...0.5]$ and $y \in [0...0.5]$ (in fractional coordinates of the surface unit cell with $a_{100}^{GGA} =$ 2.79 Å) was divided into 4 parts in every direction. The resolution in z direction was 0.1 Å in the interval of adsorption heights from $z \in [0.9...3.2]$ Å above the surface. This results in

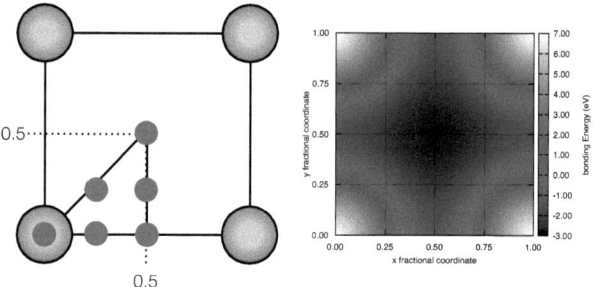

Figure 9.10: The periodic surface potential for the bond energy of a single Cu atom to the Ag(100) substrate was generated based on DFT singlepoint energy evaluations on a grid of 4 intermediate points dividing the fractional coordinate of 0 and 0.5 of the $p(1 \times 1)$ unit cell of the silver surface in x and y direction (left side: only some points are given for illustrative purposes). All necessary combinations of these intermediate coordinates were calculated, the whole unit cell potential was generated by symmetry operations afterwards. This scan was performed for different adsorption heights between a z coordinate of 0.9 Å and 3.2 Å in steps of 0.1 Å. The resulting three dimensional grid representation of the PES was then used to parametrize a $3D$ spline interpolation. Shown on the right side is one resulting $2D$ cut through this PES for an adsorption height of 1.7 Å above the surface.

9.3. Theoretical Investigation

given by

$$E^{\text{bind}}(\text{Cu@Ag}(100)) = E^{\text{total}}_{\text{Cu@Ag}(100)} - E^{\text{total}}_{\text{Cu}} - E^{\text{total}}_{\text{Ag}(100)} \quad . \tag{9.6}$$

These data points were then used to setup a 3D spline interpolation of the resulting binding energy as a function of the coordinates, which was used in the global optimization routine.

Copper Interaction Pair Potential For the parameterization of the pair potential a copper dimer was placed in a large vacuum box separated by a distance r. The resulting binding energy was calculated as

$$E^{\text{bind-pair}}_{\text{Cu}}(r) = E^{\text{total}}_{\text{Cu}_2}(r) - 2 \cdot E^{\text{total}}_{\text{Cu}} \quad . \tag{9.7}$$

The resulting data points were used as input for a least-squares fit of a Morse potential of the form:

$$V(r) = D_e \cdot [1 - \exp(-a(r - r_e))]^2 - D_e \quad . \tag{9.8}$$

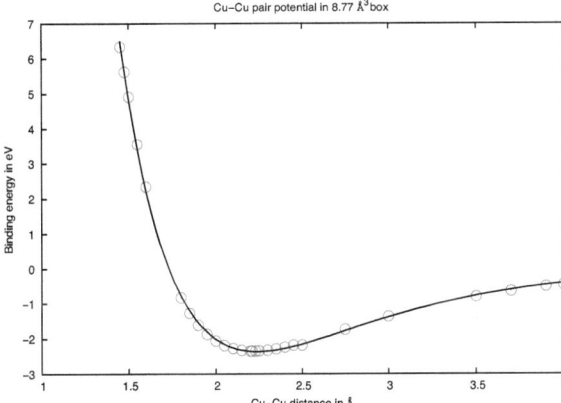

Figure 9.11: Data points used for the parameterization of a Morse potential to describe the pair interaction of copper atoms. Red circles are DFT data points, the solid line represents the resulting Morse fit.

The resulting set of parameters D_e, r_e and a is given by: $D_e = 2.367$, $a = 1.364$ and $r_e = 2.240$. It should be noted, that the equilibrium distance, r_e between the copper atoms in vacuum is not the same as the separation in the bulk ($r_e = a_{100}^{\text{GGA}} = 2.564$ Å). The vacuum equilibrium distance is shorter, which is of course due to the fact that there is only one binding partner available for the artificial vacuum dimer. However, one can argue that this stronger copper interaction could compensate the lack of many-body interaction terms describing the adsorbate–adsorbate interactions within the Frenkel-Kontorova description. Another point noteworthy is the fact that on a surface the adsorbate–adsorbate interaction naturally gets screened by the interaction of adsorbate species with the surface potential. This effect is obviously not covered in the vacuum pair potential. The results of this screening are a weaker adsorbate–adsorbate interaction and a shift in the equillibrium distance. Of course, these two effects are correlated. To cover this in the parameterization of the pair potential is not trivial. The question arising is consequently how to separate the contributions to the total first priniples energy of adsorbate–substrate interaction from those of the adsorbate–adsorbate interaction. An attempt to this problem has been made, but the resulting potential curve did not fit the classical Morse potential description so well anymore. Moreover, using this screened pair potential based fit or the stronger adsorbate–adsorbate interaction from the vaccum pair potential did not change the qualitative results after the global optimization attempts. Therefore, here only results employing the idealized vacuum pair potential for the description of the Cu–Cu interaction will be presented.

Global Optimization by Simulated Annealing - Searching for Better Structural Motifs

In addition to the definition of the so-called *cost function*, several other parameters have to be set in a simulated annealing run. In the global optimization problem that will be tackled here with the method of simulated annealing the cost function is of course the total energy of the system as given by the parameterized Frenkel-Kontorova model for this system. Therefore, the criterion to accept a newly generated structure within the optimization is obviously, if that new structure presents a lower energy than the current structure.

A simulated annealing run can be described in a simplified view as a sequence of Monte Carlo Simulations in the canonical ensemble (N, V, T – fixed), where the temperature T is quenched by a certain cooling rate from a starting temperature to a final temperature. All these variables are representing the additional set of parameters, that have to be specified additionally to the definition of the cost function. One has to define a starting temperature and a final temperature, a cooling rate and a number of Monte Carlo steps to be performed before the next temperature value is sampled or to define another criterion that specifies this. In the optimization attempts for the island structures, the starting temperature was always chosen to be 1000 K and the final temperature was selected to be in the range of the reported temperature of the STM experiments, namely $T_{\text{final}} = 250$ K. As cooling rate a fixed ratio of $c = 0.995$ was chosen and the new temperature was generated by $T_{i+1} = T_i \cdot c$. For every temperature 1000 Monte Carlo steps were performed. This choice may seem rather aggressive, however, one has to take into account that the system sizes for which those optimization tests have been performed were rather small (see discussion below). Moreover, the dominant deficiency of the cost function definition that has been revealed by these optimization tests would not have been affected by a higher number of Monte Carlo sampling steps in particular. In general, finding the optimal annealing scheme for a certain optimization task can be quite cumbersome, is not always a straightforward task and may be different for different systems.[187] Since the focus of this work was not in particular placed in the regime

9.3. Theoretical Investigation

of finding the global minimum structure of all possible island structure, but rather focused on gaining insight about new structural motifs and the stability of the initially guessed DFT input structures, those choices were once made and not subject to optimization attempts afterwards anymore.

Finally, a displacement procedure for the island atoms must be defined, for which a random displacement of particles within an active volume above the surface has been selected. The boundary conditions of this active volume were on the one hand determined by the (x, y) area of the surface that was used in the input structure description. It was ensured that this area exceeded the island dimensions of the DFT structure after relaxation that had the same number of Cu atoms of about a minimum expansion factor of 4. The boundary conditions for the z displacement were on the other hand determined by the range of z values for which the spline interpolation of the surface potential had been parameterized. This z range was in turn selected based on the adsorption heights found after the DFT geometry relaxations and then expanded about 2 Å into the vacuum.

«Conservative» Frenkel-Kontorova Modeling of a 3D Structure

As a first step to gain insight into the reconstruction behaviour of a certain island size from the point of view of global optimization, the starting configuration for the optimization run was chosen to be the reconstructed final configuration of the (6 × 6) island geometry optimization from a DFT calculation. The structural information and all necessary run time parameters and potential parameters were provided to a self-developed simulated annealing code written in C++ in form of flexible text based configuration files. In figure 9.12 the final configuration for the target temperature is shown. Two things are worth notic-

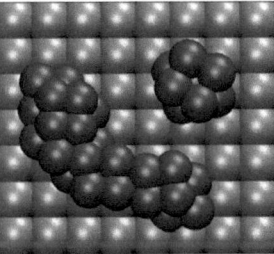

Figure 9.12: Annealing attempt for a (6 × 6) island struture. The starting structure was the relaxed DFT structure shown in Fig. 9.17.

ing about this final structure: (i) in fact, hexagonal (111)-like facets can be seen for the copper atoms, similar to the final structure from the DFT calculation. However, and more remarkably, (ii) the island structure has been completely torn apart and is not a compact island structure anymore. Several different starting structures have been tried to optimize within this extended Frenkel-Kontorova description presented here and for all tryouts the final structures were showing these two mentioned remarkable features. Therefore, only one optimization attempt was shown here as an example.

In a first interpretation, we attributed in particular the fact that no compact island structre was retained during the optimization to the quite restricted and poor description of the Cu–Cu interactions within the island. The pair potential evaluation in the cost function algorithm up to this point followed the conservative scheme of the one-dimensional Frenkel-Kontorova model. Within this scheme, only next neighbors are taken into account. In our extension to a 3D space, bonding between Cu atoms was only evalutated if the distance between the two corresponding atoms was below a hard-coded cut off radius.

As evident by the dissolved island structure, this description cannot grasp the appropriate attractive interaction between all island atoms to allow for compact island structures. An important point to note here is that this conclusion drawn from this observation is independent from answering the question whether the simulated annealing parameters have been appropriately set to find the global minimum structure. Even for the case that by wrongly chosen parameters the optimization procedure would have been trapped in a local minimum, the total energy of the resulting dissolved island structure is lower than the one of the starting structure within the Frenkel-Kontorova energy description. This, however, is inconsistent with the experimental data, in which no split island structures have been observed. We conclude therefore from this observation that the potential description is just not sufficient to cover the basic physics behind the island formation and reconstruction in the current setup.

A Step Further: Taking Long-Range Pair Interactions into Account

A quite straightforward attempt to extend the potential description used in the paragraph before is to eliminate the condition of a cut off radius from the pair potential evaluation. This can be easily done in the program implementation and it means that now also the weakly attractive long tail of the pair potential is taken into account for any configuration of Cu atoms. However, it should be stressed that also within this description still the attractive interaction of island atoms is purely modeled by a pair potential. No many body effects are taken into account. To further

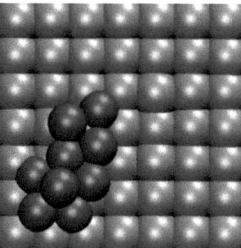

Figure 9.13: The final structure for a (3 × 3) island as coming out from a simulated annealing attempt using a Frenkel-Kontorova model with long-range pair potential.

eliminate a possible negative influence by an unsufficient Monte Carlo sampling within the simulated annealing steps, the search space was reduced by reducing the number of island atoms from a (6 × 6) island to a structure with 9 atoms in the simulation cell. The initial configuration of the island atoms consisted of randomly distributed atoms within the active volume to investigate in particular the attractive interaction between the island atoms. A successful description would be indicated by the formation of a compact island structure within the simulation. Obviously, the extended pair potential description allowed this time the formation of a compact island structure. In contrast to the previous result this is an im-

9.3. Theoretical Investigation

provement in the description. However, another noteworthy feature of the island structure is the fact that due to the isotropic pair potential the formation of three-dimensional almost chain like structures of spherical sub-structures. In which there is one circular shell of atoms in the nearest neighbor distance of a central atom. This configuration is thus apparently preferred within the employed isotropic pair potential description. It is perfectly clear, that in a configuration like that the highest number of neighbor atoms with a distance very close to the distance of the deep well in the pair potential can be realized. However, comparing to the experimentally found structures, which are mono-layer configurations or at most two-layer configurations the formation of spherical cluster structures adsorbed on the surface seems not to represent the true situation.

In an interpretation of this result, this failure is again contributed to the missing many body interactions in the purely pair potential based Frenkel-Kontorova parameterization. In addition, the many body interactions between island atoms do not necessarily have to be isotropic and therefore may be the underlying reason for the island structures observed in experiment. These shortcomings of the interaction description between island atoms laid the foundation for the decision for another potential description of this system for further investigation.

The potential description that was chosen is the embedded atom method (EAM) description. As already stated in the introductory chapter of the methods used, the evaluation of an EAM configuration energy is only fairly more expensive than the evaluation of a pair potential. However, by means of the embedding term and by a parameterization to a set of experimentaly known values of certain properties, many body interactions are accounted for implicitly.

9.3.5 An EAM Potential Description of the Energetics

Motivated by the simulation attempts described above, we switched to the semi-empirical embedded-atom potential description for the Cu/Ag system.[107] All simulations were performed using the LAMMPS MD code[188] and the functionalities of local minimization procedures implemented in it (steepest descent and conjugate gradient algorithms as described in section 7.1). After switching the potential description, a first concern must go to the validation of the energetics predicted by the new potential against the set of known energies calculated with the higher level method (DFT).

Therefore, all island structures formerly calculated with DFT are recalculated with the EAM potential and the resulting stabilization energies are compared. Careful attention was paid to the fact that within the EAM parameterization a database including experimental lattice constants for Ag and Cu had been used, cf. chapter 6. Consequently the final structure of the GGA-PBE calculations were scaled by a factor resulting from the ratio between the equilibrium lattice constants as predicted by the GGA-PBE calculations and as known from experiment. Fig. 9.14 shows the resulting stabilization energies. We find the EAM E values to be in good agreement to the DFT values. Furthermore, the trend in the island stabilization is nicely reproduced. We attribute the constant offset in the E to the LDA parameterization used for the Cu–Ag interaction in the EAM potential (see chapter 6). The LDA xc-functional is known to overestimate the interfacial binding and therefore this potential favors the all-hollow islands with the larger contributions by interfacial binding between Cu and Ag (corresponding to a smaller value of the stabilization energy).

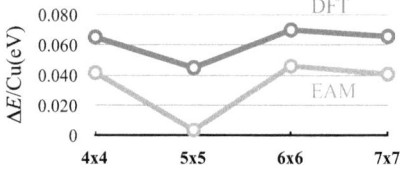

Figure 9.14: Comparison between stabilization energies E(eV/Cu) as predicted by DFT-GGA calculations and predicted within the EAM potential. The trend of the stabilization energy with island size is reproduced nicely. A constant offset is caused by the LDA dataset used for the alloy-part parameterization in the EAM potential.

Having a validated and, compared to DFT, numerically undemanding potential for our system, we then tried to estimate the critical island size for the stability turnover. For this we followed the same approach as for the DFT calculations. We started with two different initial structures and relaxed them locally by an energy minimization procedure. Then we compared the final energies and plot the energy difference. By this procedure, we were able to estimate a critical island size of about (10 10), which corresponds to an approximate island area of about $8.4\,\mathrm{nm}^2$, cf. Fig. 9.15. This island size is in very good agreement with the size that was observed in the experiments for the change in the island diffusion and decay properties ($7\,\mathrm{nm}^2$).

9.3. Theoretical Investigation

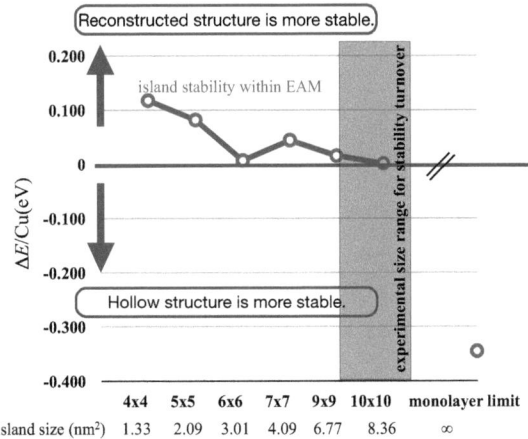

Figure 9.15: ΔE for islands up to a size of (10 × 10), which is the critical size where the stability line is crossed the first time. This size range is in excellent agreement with the experimentally reported island size of $A \approx 7\,\text{nm}^2$ in [19].

9.3.6 Searching for New Island Shapes

Finding structural motifs

In a next step, we deviated from the assumption of a square island structure and started an extensive structure search for other island shapes of the (6 × 6) island structure as working example. For this we followed a pretty simple procedure, first proposed by Pickard and Needs[120] as presented in the methods chapter, see section 7.2.2.

A given number of Cu atoms is placed randomly in a predefined volume area above the surface plane, followed by energy miminization. The final minimum energy structure is stored with its corresponding energy and the next structure is generated. We defined the accessible volume for the initial placement of Cu atoms to be centered in the unit cell being half of the unit cell size in x and y direction, and the chosen z coordinate of a Cu placement not to exceed the height of a double layer of Cu atoms, initially. The succesive minimization was unbiased. By this procedure, we were able to scan about 10^6 structural motifs for this single island size with their tabulated EAM energies.

DFT Refinement of Favorable Island Shapes

Following this search step, we took the 5 most favorable (6 × 6) structures found in this EAM potential based structure search and re-optimized the geometries by subsequent DFT geometry optimizations starting from the found structure. Since we concentrated in this search on

(6 × 6) islands, DFT calculations were still possible. For these calculations, we used the same calculation setup as described above, with the only modification of appropriately enlarging the unit cell in z direction to account for the increased height of the island structure and ensuring again at least 6 Å vacuum between periodic images in z direction.

Figure 9.16: Most stable (6 × 6) structure, suggested by random structure search and DFT refinement.

The most stable geometry, cf. Fig. 9.16, turned out to be a compact two layer arrangement of the 36 available Cu atoms, with the first layer Cu atoms residing in hollow sites and the toplayer Cu atoms sitting in the hollow sites of the first Cu layer, pushing the second layer Cu atoms accordingly to the top sites of the underlying Ag(100) lattice. The toplayer Cu atoms have a distance of about 3.2 Å from the underlying Ag(100) surface in the relaxed geometry. In this geometry, the stabilization energy compared to the all-hollow configuration increases to about +0.135 eV/Cu. This is about twice the value for the monolayer configuration (+0.070 eV/Cu) that has been found as local minimum by the pure geometry relaxation starting from the all-bridge situation before. From this we conclude, that the assumption of Cu islands being square and exhibit only monolayer heights requires some closer investigation. Apparently, double layer configurations represent even deeper lying local minima of the Cu island structures in this surface. Moreover, recalling the fact that the experimental conditions of the STM measurements included elevated (room) temperatures, thermal activation energy will be present under the experimental conditions. As a consequence, the system will be in the configuration determined by the statistical thermal equilibrium, even if the reconstruction to form this configuration would require the overcoming of an activation barrier.

9.3.7 The Stability of Monolayer Islands

Therefore, based on the results of the extended structure search, some interest in the temperature dependent stability of monolayer island structures was raised. The experimental preparation and the final STM measurement of the island structures were performed at el-

9.3. Theoretical Investigation

evated temperatures. All stability considerations so far are based on calculations for zero Kelvin temperature and therefore neglect the effects of temperature completely. Clearly, temperature will have effects on establishing a certain statistical ensemble average and consequently the most likely configuration in thermal equilibrium.

Starting from the assumption that initially monolayer island structures are formed under the experimental conditions, the formation of double layer islands would involve re-ordering or «hopping» events to transport atoms on top of the island structure. This assumption is rationalized by the fact that the experimental conditions involved sub-monolayer coverages build up by vapor deposition techniques, where usually single Cu atoms are adsorbed on the surface and then clusters are formed by diffusion of those single atoms. Obviously, the formation of a monolayer structure in the initial phase is the most likely one in this scenario. The formation of the second layer would then happen by Cu atoms rearrangements. Particularly prone for hopping events are the atoms located at the rim of the island structure. This assumption can be justified by a simple bond-order argument. The cost to break the bonds of a rim atom is smaller than the one needed to extract an atom from the center of an island structure. A centered atom has got much more neighboring atoms to which it is bonded to.

Nudged-Elastic Band Calculations for a First Atom Hopping Process

In order to get a first idea of the energy that is needed for the first hop of a corner atom on top of the island a series of NEB calculations were performed using the potential description provided by the EAM potential. In all these calculations the initial structure was the (6 × 6) reconstructed island structure shown in figure 9.17. The final configuration for the NEB minimum energy path search setup was generated by lifting one selected atom to a position on top of the island. Between initial structure and final structure 5 images were generated by a simple interpolation of coordinates. During the NEB optimization the positions of all island atoms except the one supposed to hop on top of the island were frozen. Obviously, this setup represents a quite artificial situation. However, these constraints were necessary to allow the NEB algorithm to converge to a minimum energy path.

Figure 9.17: The relaxed and reconstructed island structure of a (6 × 6) island that was used as initial structure in all investigations connected to the question of the stability of monolayer islands.

Despite this clear shortcoming, it should be noted that these calculations were not meant to necessarily provide the true reaction path for a hopping process. The intention behind these calculations was more to get an order of magnitude for the hopping process to gain some rough insight into the probability of double layer formation from monolayer islands under the experimental conditions, in particular to gain an estimate about the influence of thermal activation.

For all tested hopping moves, the calculated barrier of the energy path as resulting from the NEB optimizations was between 1 eV and 1.5 eV. These values can be taken as an upper bound for the activation barrier of a hopping process and the true activation barrier will most likely be lower. The reason for this argument is given by the constraints employed in the NEB optimization. All island atoms except the hopping one were constrained in their movements and allowing a relaxation for all island atoms would most likely have lead to lowering of the final energy. Based on the Arrhenius equation a rough estimate for a critical temperature for this process to happen can be estimated from this estimated barrier.[189,190]

It is $T \approx -\Delta E \cdot \ln(\Gamma_0/\Gamma) \cdot k_B$. For generic choices of surface diffusion processes of $\Gamma_0 =$ 1 THz and $\Gamma = 1$ Hz and ΔE chosen to equal 1 eV (being the lowest energy barrier found in the NEB optimizations), we get a predicted temperature, T, around 400 K for which a hopping would occur with a frequency of one event per second. This crude estimate, however, allows to conclude that the elevated temperatures of the experimental conditions ($T \approx 250 -$ 300 K) are close to the temperature range for which an activation of hopping processes would have to be expected. Therefore, based on this first theoretical results, the necessity for high temperature experiments to specifically scan for temperature effects can be concluded.

Parallel Replica Dynamics – MD Simulations at Elevated Temperatures

As a second theoretical approach to study temperature effects on the stability of the monolayer (6 × 6) island shown in figure 9.17 two configurations were investigated using accelerated molecular dynamics[135] simulations, again using the EAM potential description for the system's energetics. In each case, the system was initially prepared with a kinetic energy distribution that resembled a certain target temperature. After this, parallel replica dynamics simulations were performed in the microcanonical ensemble. Every 250 fs steepest descent minimizations were performed to check for diffusion or hopping events of island atoms. The criterion to detect such an event was given by a displacement of a Cu atom of more than 2.9 Å ($= a_{100}(Ag)/\sqrt{2}$)).

The resulting structures of those simulations are shown in figure 9.18. In an interpretation of the structural evolution after a cumulated simulation time over all replicas of 60 ns it seems fairly safe to state that elevated temperatures close to the experimental temperature range strongly promote the formation of double layer island structures. This double layer formation is a fast process which is almost completely finished already on the very short time scale of the performed molecular dynamics simulation. This does not exclude the formation of double layer structures on longer time scales for lower temperatures, of course. However, this result again points to the need for high temperature measurements to thoroughly study whether double layer structures can or cannot be seen for certain experimental temperatures.

9.4. Conclusions and Outlook

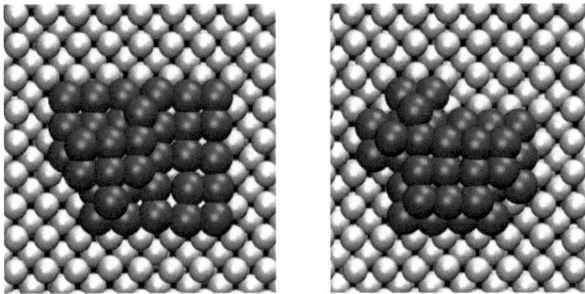

Figure 9.18: Resulting island structures for PRD MD runs. The simulations were performed in the NVE ensemble with an initialized kinetic energy distribution that resembled a certain target temperature. Both structures were obtained after a total cumulated time of about 60 ns in the PRD formalism. (left side) Final structure for a initial kinetic energy distribution of all atoms in the simulation cell representing a temperature of 200 K.
(right side) Final structure for a initial settings in order to represent 300 K.
In the comparison, it is obvious that an increased temperature favors double layer formation of islands.

9.4 Conclusions and Outlook

In conclusion, the binary system of Cu islands in the sub-monolayer coverage regime at the Ag(100) surface provides a rich variety of structures. Motivated by experimental data we were able to address this structural complexity in a multi-scale modeling approach. To successfully achieve this, a proper description of many-body lateral adsorbate–adsorbate interactions turned out to be essential. We were able to cover these effects by an EAM alloy potential, parameterized for this system and validated against a set of structures for which first-principles energies were known from density-functional calculations.

Within this approach, we could identify and explain the physical reason for the size-dependent reconstruction behavior of monolayer island structures seen in the experiments. The reconstruction is driven by the interplay presented by the adsorbate–substrate bonding and the adsorbate–adsorbate bonding (strain release). This is caused by the rather strong lattice mismatch between the involved materials.

The additional question that was raised throughout this investigation was whether the assumption of all islands being monolayer structures represents a reasonable one for the working conditions of the experiments. This was answered with accelerated molecular dynamics simulations for this system. As a conclusion drawn from our theoretical results a clear need for additional experiments can be postulated, since the temperature range of the current set of experiments are in very close proximity to a region where thermal activation for the double layer formation might in fact play a role. It is also not uncommon for metals that are immiscible in the bulk to allow for a certain alloying in the situation where the alloy is confined on a surface. This has been proposed for the system of silver growth on copper[191] and this assumption would present a next step to be checked for. Also theoretical investigations are available for this particular situation of alloy formation, as already mentioned in the introduction to this chapter[176].

Another subject of interest would be to compare the diffusion constant derived from simulations to the experimentally measured diffusion constants for the different islands and finally identify a mechanism for the surprisingly fast diffusion of small islands. We could contribute to this question by means of a first rationalization, namely by comparing the bond strength to the Ag(100) surface of different island sizes and by observing a weaker bonding of small island structures. A deeper investigation of this would definitively be an interesting but also challenging task and could present one focus of future investigations.

In addition, the possibility of a surface confined alloy formation is a point which needs further investigation. It is not uncommon for metals which are immiscible in the bulk phase to form a stable alloy in the first few layers of the surface. From a chemical point of few, this opens some interesting aspects with respect to new catalytic properties which might be presented by these surface alloy phases.

Part IV

Conclusions

Chapter 10
Final Remarks and Future Steps

fama di loro il mondo esser non lassa;
misericordia e giustizia li sdegna:
non ragioniam di lor, ma guarda e passa.

Dante Alighieri

In this work two methods for the description of lateral interactions between particles adsorbed on surfaces have been examined. One approach was based on an expansion of the particle interaction in terms of clusters of interactions, which were then assigned to certain energy contributions within a parameterization process from first principles. With the final set of parameters arbitrary adsorbate configurations up to half a monolayer coverage could be described to a high accuracy close to the level of first-principles calculations for negligible computational cost. For higher coverages, however, the inability to properly account for axial bending of the adsorbed CO particles as a result of local geometry relaxation lead to a break down of the static cluster expansion approach taken here. This was clearly visible in the ground-state validation procedure. On the one hand, it is well known that not always the cluster expansion with the best CV score is able to predict the correct ground state structures. In these case, one approach can be to include different configurations with different statistical weights. This would be a step for further improvement of the cluster expansion presented here [95,96]. On the other hand, an improved description that would allow for more flexibility could also present a direction for further investigations. One method to mention here would be the *connector-model* proposed by Tiwary and Fichthorn [192].

The second limiting case for the relation between the two influencing interaction strengths was studied choosing a different methodological approach. An established semi-empirical potential description was used that includes the effects of lateral interactions between the particles indirectly. The parameterization of this potential included experimental data and data from first-principles which indirectly covered the effects of these interactions. The question whether this potential was then also able to cover the dominating effects causing the effects that were studied within this thesis was answered with a validation against data from density-functional theory calculations. In this way, a size regime close to the experimental one was accessible by our simulations. An explanation of the observed reconstruction behavior was presented based on the microscopic insight that were gained from our simulations.

The importance of lateral interaction for either ordering behavior or structure formation of adsorbate islands has been become clear from the two example systems. These two examples show clearly that a proper description of the interactions between adsorbates is a necessary requirement for all meaningful mesoscale model descriptions. The work presented so far showed that a consistent derivation of these parameters is possible from first-principles calculations directly in the form of the cluster expansion approach or indirectly by means of the fitting procedure employed in the parameterization of the embedded-atom-method potential and a subsequent validation of its applicability for the specific question.

10.1 Future Steps

Once parameterized the beauty of the model descriptions developed here lies in their small demands — compared to electronic structure methods — of computational power for the evaluation. Therefore, accessing the mesoscopic regime and evaluating real statistical averages of thermodynamical ensembles becomes feasible and allows the prediction of material behavior under realistic conditions.

For this reason future steps to be taken from the final stage reached within this thesis would be presented by studies about the behavior of the example systems presented here on the mesoscale. For the adsorption of CO on Pd(100) studying ordering behavior for increasing coverages at realistic temperature ranges is now possible and would allow direct comparison to experiments.

For the island growth of Cu islands on the Ag(100) surface a challenging future step is presented by the theoretical determination of the diffusion constants and the identification of the diffusion mechanism of small islands. Even with the semi-empirical potential description taken here this presents still a challenging task. Again, a favorable result of such an investigation would be presented by the possibility to compare to experimental results directly.

Appendix A
Convergence Tests

All simulations which are based on numerical methods and the self-consistent evaluation of quantities should follow a two step approach to define relevant parameters for the performed calculations.

At first, the quantity of interest should be identified. For example, this can be a certain material property or energy value. In the second step, the behavior of this quantity with respect to the relevant variables of the calculation method must be investigated and a certain set of parameters should be identified with provides the interesting quantity to a defined precision for an at best minimal computational cost. Those systematic identification of the calculation parameters is known as convergence tests.

In this chapter the results for the convergence tests of this work will be presented. After the quantity of interest has been described, the results of the convergence tests and the final set of calculation parameters will be presented for each studied case.

A.1 Convergence Tests for the DFT Calculations

In the following section the systematic convergence tests performed for the the different scenarios for which DFT energies have been calculated are presented. The convergence behavior of the respective quantity of interest will be checked with the most important parameter settings for the DFT calculations using an expansion in plane-waves and pseudo-potentials. These are namely, the maximum lattice vector G and the density of the k-point mesh used for the evaluation, see section 3.3 for more details on these two quantities.

A detailed description of the convergence tests will be presented here for the system which was taken as showcase for the cluster expansion approach for the description of lateral interactions, the adsorption of CO on Pd(100), chapter 8. For the second system, investigated throughout chapter 9, well converged settings were taken from a former work by McNellis published in [166]. The DFT program package (CASTEP[74]) used for all DFT calculations of the work presented here had also been used in the work of McNellis.

A.1.1 Bulk Properties

The first quantity of interest is the GGA-PBE[60] equilibrium bulk lattice constant. In addition to the xc-functional, the value of the lattice constant will depend on the specific parameter settings for the calculation and its value will deviate from the experimental equilibrium lattice constant. For the determination of the theoretical equilibrium lattice constant bulk energy values for several different lattice constants (ideally around the equilibrium value) are calculated and the resulting energy-volume curve, $E(V)$, is fitted to the Murnaghan equation of state[193]:

$$E(V) = E_0 + \frac{B_0 V}{B_0'}\left(\frac{(V_0/V)^{B_0'}}{B_0'-1}+1\right) - \frac{B_0 V_0}{B_0'-1} \quad . \tag{A.1}$$

This determination is performed for several different settings for the plane-wave expansion and the k-point sampling. If the target quantity does not change its value more than a threshold value anymore, it is said, that the quantity is converged with the corresponding settings.

Bulk Palladium, Pd, crystalizes in a face-centered cubic structure with an experimental lattice constant in the bulk of $a_0(\text{Pd}) = 3.880$ Å[194], and a bulk modulus of $B_0(\text{Pd}) = 180\,\text{GPa}$[177].

Figure A.1 shows an example of a resulting $E(V)$ curve for one combination of plane-wave expansion and k-point sampling.

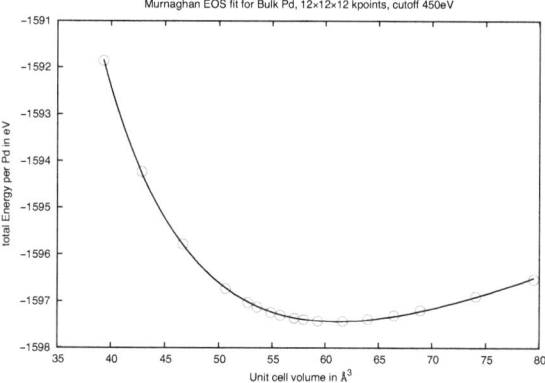

Figure A.1: $E(V)$ curve for the palladium bulk structure and the corresponding fit by the Murnaghan equation of state

As explained already, similar plots were performed for several different settings and the resulting development of the so-determined theoretical equilibrium bulk lattice constants is shown in Fig. A.2

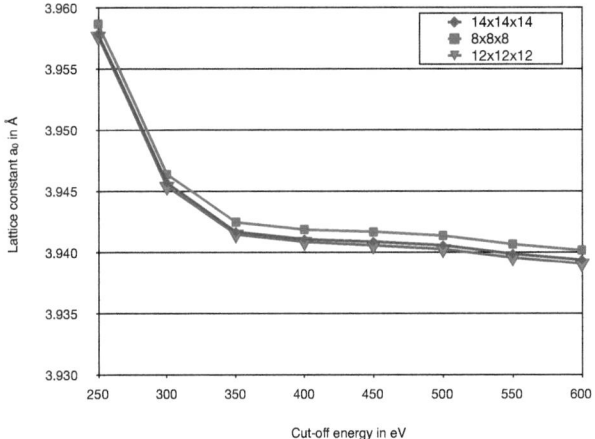

Figure A.2: Convergence behavior of bulk lattice constant of Pd with respect to different values for the cut off energy for three different integration grids. Plotted are the results of Murnarghan fits performed for the given parameter combination of cut off energy and k-point sampling

It can be seen that the value for the theoretical lattice constant is changing less than ± 0.0003 Å for an increase of the cut-off value from 450 eV to 500 eV anymore (for all plotted k-point grids). In addition, the difference between a k-point sampling grid of $8 \times 8 \times 8$ and a finer grid of $14 \times 14 \times 14$ is in the order of ± 0.0008 Å. Therefore, the choice for the final settings for these two parameters has been a value of $E^{\text{cut-off}} = 450$ eV and a k-mesh of $8 \times 8 \times 8$ for the conventional unit cell of palladium.

These settings yield a GGE-PBE equilibrium lattice constant of 3.942 Å which is in close agreement to the previously reported all-electron result of 3.95 Å. Also the value for the bulk modulus $B_0 = 1.54$ Mbar agrees well with the value reported ealier ($B_0 = 1.57$ Mbar)[18,195].

A.1.2 Checking the Quality of the Reported Parameter Settings for Silver and Copper

Bulk Silver, Ag, crystallizes in a face-centered-cubic structure with an experimental lattice constant of $a_0(\text{Ag}) = 4.085$ Å[196] and a bulk modulus of $B_0(\text{Ag}) = 100$ GPa[177].

For the reported[166] parameter set of a $12 \times 12 \times 12$ Monkhorst-Pack integration grid of the Brillouin zone and a plane wave cut off energy of $E_{\text{cut}} = 400$ eV for the bulk structure of silver, a set of bulk energies for different lattice parameters have been calculated and the resulting $E(V)$ curve, see Fig. A.3, was fitted to Eq. A.1 like for palladium before. In this way, the theoretical equilibrium lattice constant was determined and compared to previously reported values.

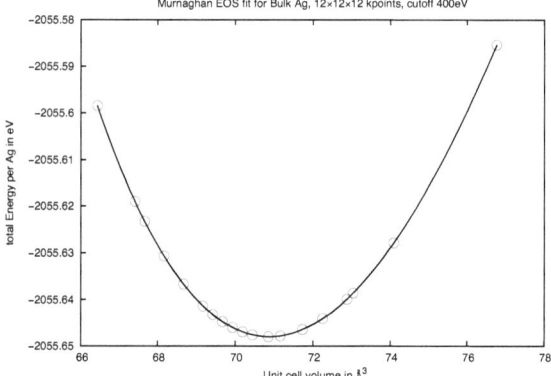

Figure A.3: Fit result for the $E(V)$ curve of bulk fcc structure of silver with the PBE xc functional.

The result of this process is a silver equilibrium lattice constant for the GGA-PBE functional of $a_0^{\text{PBE}}(\text{Ag}) = 4.1377$ Å, which is the lattice constant used for all calculations and data evaluations within this thesis. Another important quantity to measure the quality of the calculations is the bulk modulus, B_0. The fit produces a value of $B_0 = 0.98$ Mbar. Both predicted values agree very well with what is found in the literature for this functional.[197]

For the convergence behavior of the copper bulk lattice constant results of identical quality have been obtained with the parameter setup used for silver. Therefore, only the silver results have been presented here in detail. The fitted PBE lattice constant of copper was found to be $a_0^{\text{PBE}}(\text{Cu}) = 3.6258$ Å and a $B_0 = 1.45$ Mbar which is as well in excellent agreement with previously reported results[198].

A.1.3 Surface Properties of the (100) Surface of Palladium

New variables are introduced into the calculation setup by the attempt to describe a surface within a $3D$ periodic code by a supercell approach. On the one hand, the surface must be described properly and interaction through the vacuum with the next periodic image along the surface normal must be avoided. On the other hand, the slab must not only describe the surface, but also the bulk of the material properly. To achieve this, a common approach is

A.1. Convergence Tests for the DFT Calculations

to keep the atoms on one side of the slab fixed to the bulk positions by imposing a constrain in the calculation setup. In addition to these new parameters, also the convergence with respect to the cut-off energy and the integration grid has to be checked again for the new scenario.

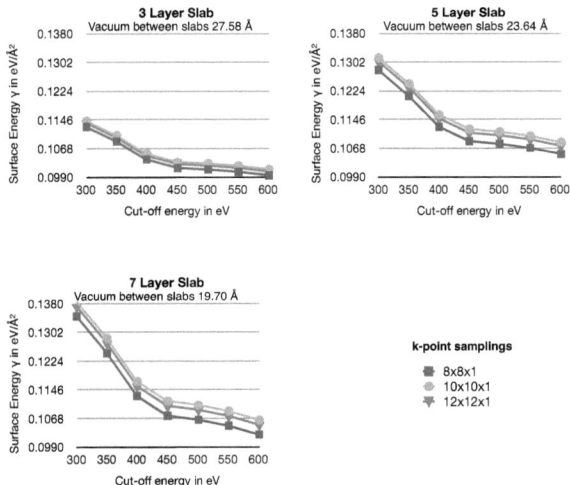

Figure A.4: The convergence of the surface energy γ for different settings of the cut-off energy parameter and the k-point meshes for three different slab sizes and vacuum separations.

The quantity of interest which needs to be checked for convergence here is the surface energy, γ as defined in subsection 2.2.2. The surface energy has been determined for three different setups of the unit cell. Setups for a three layer slab of palladium atoms with a vacuum separation of 27.58 Å, a five layer slab with a separation of 23.64 Å and a seven layer slab with 19.70 Å of vacuum between the periodic images were generated. For each of these setups three different sampling grids in k-space and seven different values for the cut-off energy were used. the two top layer of palladium atoms were always allowed to relax in the performed geometry relaxations, the remaining bottom layer(s) were kept fixed to ideal bulk positions using the equilibrium bulk lattice constant determined before. The corresponding values for the surface energies are plotted in Fig. A.4.

By this approach, a clear separation of the effects originating from the number of layers and the influence of the vacuum separation cannot be performed. It can, however, be seen that the surface energy only changes in the range of at most 10 meV throughout all calculated geometric setups for a converged pair of cut-off energy and k-point integration grid, as shown

130   Appendix A. Convergence Tests

layers	vacuum separation in Å	surface energy γ in eV/Å^2
3	27.58	0.1017
5	23.64	0.1098
7	19.70	0.1083

Table A.1: The surface energy of a Pd(100) surface as calculated for several slab thicknesses in a $p(1 \times 1)$ unit cell.

in an example in table A.1 for a cut-off energy of 450 eV and a k-point sampling grid of $12 \times 12 \times 1$. The calculated energies are also in good agreement to the literature reference values of all-electron calculations[198].

A.1.4 The Gas-Phase Reference State for CO

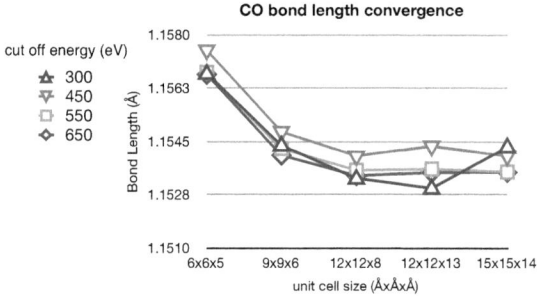

Figure A.5: Shown in these graphs are the convergence tests that have been performed for the gas phase reference of CO. The relaxed bondlength is plotted against the unit cell size for several values of the cut-off energy. All calculations have been performed with a Γ-point sampling of the Brillouin zone.

For the calculation of binding energies (see for a definition 2.2.3) a reference energy for the gas state of the adsorbate needs to be defined. In the case of carbon monoxide (CO) the reference state can be conveniently selected to be the energy in the gas phase of a single free molecule. This decision, however, imposes a challenge for the usage of a periodic density-functional theory code to calculate the electronic total energies. The CO molecule is placed

A.1. Convergence Tests for the DFT Calculations   131

in a periodically repeated box of vacuum, defined by the unit cell parameters. To avoid artificial influences by an interaction of the periodic images, it has to be checked, if the vacuum separation is large enough not to influence the energetics of the isolated molecule anymore by subsequently enlarging the box and checking for the resulting change.

The determining quantity which needs to be checked for convergence here is the CO bondlength. Since the resulting unit cells are large the integration was performed only at the Γ-point in all convergence test calculations presented in the following. A test calculation using a finer k-point grid (the one found to produce converged results for the surface and slab geometries) did not change the resulting converged bond length. Another important aspect is posed by the usage of pseudopotentials. It is well known that for the description of oxygen in particular rather high cut off energies are necessary. For this reason the cut off values for the total energy reference of the gas phase CO are chosen rather high. The results of these calculations are plotted in Fig. A.5.

For all subsequent calculations, the gas phase reference value for a unit cell of $15 \times 15 \times 14$ Å^3 with a cut off energy of 650 eV is used. The Brillouin zone is sampled at the Γ-point. With these calculation parameters, the reference energy is given by $E_{tot}(CO_{gas}) = -590.22$ eV and the optimized CO bondlength is converged to a value of $d(CO) = 1.153$ Å with respect to either vacuum separation or cut of energy. This value is in good agreement with the literature reference value for the PBE exchange correlation functional, see Ref. [199] and references therein.

A.1.5 Adsorption of CO on Pd(100)

For the adsorption, the binding energy of a CO molecule to the specific adsorption site as defined in subsection 2.2.3 is the quantity which needs to be checked for convergence. The corresponding parameters for the calculation are again the cut-off energy value and the density of the k-point sampling grid. Therefore, a series of calculations with several different cut-off values and three different k-point grids was performed for the three different slab geometry setups already used already to check for the convergence of the surface energy, see A.1.3. The resulting development of the binding energy is plotted in Fig. A.6. It can be easily seen that the influence of the k-point sampling grid is rather small, since all curves fall basically on-top of each other. The convergence behavior with respect to the cut-off value is less satisfying. In the end, a cut-off value of 450 eV was taken for the expansion in plane-wave with a k-point sampling grid of $12 \times 12 \times 12$. The reason for this decision was in this case mainly the computational cost connected to much higher cut-off values. In addition to these settings, table A.2 gives an overview about the variation of the calculated binding energy with respect to the three unit cell geometries used. It can be seen that the difference in the CO binding energy of a five layer setup and a seven layer setup is rather small. Again, from these three geometrical setup the separation of effects on the quantity of interest originating from the number of layers in the slab and the vacuum separation cannot be clearly separated. For such a separation individual convergence test for either of the two variables should in principle be performed.

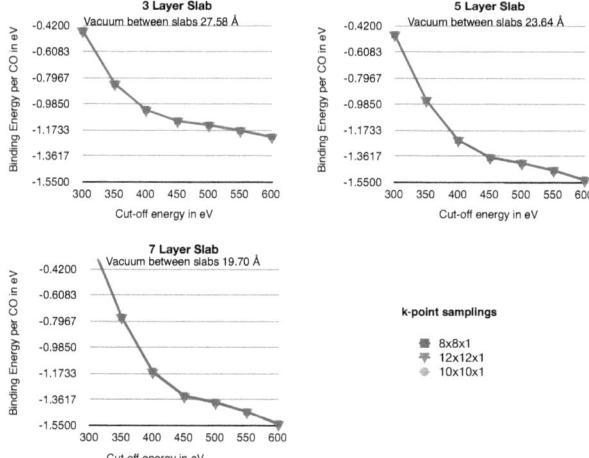

Figure A.6: The convergence of the binding energy of CO to the bridge site of the Pd(100) surface for different settings of the cut-off energy parameter and the k-point meshes for three different slab sizes and vacuum separations.

layers	vacuum separation in Å	binding energy in eV/CO
3	27.58	−1.106
5	23.64	−1.370
7	19.70	−1.379

Table A.2: The binding energy of a CO molecule on the bridge site of a Pd(100) surface as calculated for several slab thicknesses.

A.1.6 Adsorption of Cu on Ag(100)

For the DFT study of the formation of Cu islands on a Ag(100) surface, the quantity of interest was the energy difference between two island configurations. Therefore, this value was checked for two different island configurations, namely the infinitely large island represented by the periodic $p(1 \times 1)$ structure and the first island showing sufficient reconstructions for a significant stabilization energy, namely the (4×4) island within a (6×6) simulation cell.

This ΔE in meV/Cu was tested for convergence with respect to the cut-off value of the plane waves, the slab thickness and the vacuum separation by a couple of test calculations. The results of these are shown in table A.3.

calculation parameters	cell (1 × 1) - 1 Cu	cell(6 × 6) with (4 × 4) island
3layer, 2fixed, 300 eV, 6 Å vacuum	−0.4320	+0.081
3layer, 2fixed, 400 eV, 6 Å vacuum	−0.4340	+0.083
3layer, 2fixed, 400 eV, 13 Å vacuum	−0.4356	+0.084
3layer, 2fixed, 450 eV, 6 Å vacuum	−0.4340	+0.082
5layer, 2fixed, 400 eV, 6 Å vacuum	−0.4151	+0.116

Table A.3: Stabilization energy convergence for several different calculation setups.

The convergence test have been kept a bit more briefly, since for the two coinage metals used here already extensive convergence tests for the CASTEP code had been performed in the group before, see [166]. Therefore, some information about the necessary settings of the cut-off energy and the density of the k-point grid for a sufficiently accurate description of surfaces for this code was available lie it was for the bulk calculations presented before. For this reason, the calculations presented here for this system represent rather a cross-check than a thorough checking for convergence.

A.2 Final Computational Setup

Here, a quick summary of the final settings used for all DFT calculations presented in this thesis using the the periodic Ultrasoft-Pseudopotential Plane-Wave code CASTEP[74] with the GGA-PBE functional[60] to treat electron exchange and correlation is given.

A.2.1 CO on Pd(100)

For the study of CO adorption on Pd(100) all surfaces were modeled in supercell geometries with 5 layer slabs and a vacuum seperation of 23.64 Å between the periodic images. The cut-off energy for the plane wave expansion was 450 eV, and reciprocal space integrations were done on (8 × 8 × 1) k-point grids for $c(2 \times 2)$ surface unit-cells (corresponding to a $12 \times 12 \times 1$ sampling for a $p(1 \times 1)$ unit cell geometry). In the calculations of larger surface unit-cells, these grids were reduced accordingly to maintain an equivalent sampling of k-points in the reduced first Brillouin zone of larger unit cells. In all calculations the adsorbate layer and the top three layers were fully relaxed until the residual forces fell below $4 \cdot 10^{-2}$ meV·Å^{-1}/atom, whereas the bottom two layers were kept fixed on ideal bulk positions using the converged value of the theoretical lattice constant.

A.2.2 Cu islands on Ag(100)

The surfaces were modeled in supercell geometries, with three Ag(100) layer slabs separated by at least 6 Å vacuum above the last atoms of each structure. All atomic positions in the topmost Ag(100) layer and in the Cu island adsorbate layer were fully relaxed. until residual forces fell below 35 meV/Å. The cut-off energy for the plane wave expansion was 400 eV, and reciprocal space integrations were done on $(12 \times 12 \times 1)$ k-point grids for (1×1) surface unit-cells. In the calculations of larger surface unit-cells starting from the (6×6) setup, these grids were reduced to a k-point grid of $(2 \times 2 \times 1)$ which almost maintains an equivalent sampling of the Brillouin zone for the small and large unit cells and therefore reduces the error to the energetics introduced by the enlargement.

Appendix B
Adsorption Structure Database for the Cluster Expansion

As an initial database for the cluster expansion, a set of 45 different adsorption structures have been prepared and their geometries were relaxed. In a later step, this initial set was extended by 10 additional structures, which replaced some strongly distorted ones in the fitting database. The final fitting database contained 37 structures, see table 8.1. These selected structures determine what is called the «DFT-groundstate», that is the set of structures that are lowest in the calculated formation energy.

In the following all adsorbate configurations are presented in their initial state and in their final state after BFGS geometry optimization. All these calculations have been performed with the periodic Ultrasoft-Pseudopotential Plane-Wave code CASTEP[74] using the GGA-PBE functional[60] to treat electron exchange and correlation. The surfaces were modeled in supercell geometries with always 5 layer slabs representing the Pd(100) surface and a vacuum separation between periodic images in z direction of 23.64 Å. The adsorbate layer and the first two surface layers were fully relaxed in the geometry optimizations, the bottom three layer were fixed to ideal bulk positions. All other settings are given in the overview below:

Calculation Parameters

- Cut-off energy: 450 meV
- k-point sampling: $8 \times 8 \times 1$ grid for the $c(2 \times 2)$ cell, for larger unit cells the sampling grid was reduced to maintain an equivalent sampling of the Brillouin zone, eg. the $c(4 \times 3)$ unit cell was sampled with a $2 \times 4 \times 1$ grid.
- Grid scale: 1.80
- Smearing width: 0.1 meV, Hermite Polynomials
- Force convergence criteria: $\Delta F \leq 4 \cdot 10^{-2}$ eV·Å^{-1}
- Energy convergence criteria: $\Delta E \leq 2 \cdot 10^{-6}$ eV/atom
- Electronic minimization convergence criteria: $\Delta E \leq 1 \cdot 10^{-6}$ eV/atom
- XC-functional: PBE

Appendix B. Adsorption Structure Database for the Cluster Expansion

B.1 Reference Energies

B.1.1 The Clean (100) Surface Unit Cells

All surface energies given here were obtained by geometry relaxations of clean (100) surface unit cells of the given size. Therefore, the clean and relaxed surface is representing the reference state for the calculations of the binding energy.

Unit cell size	E^{total} in eV
$c(2 \times 2)$	−7983.982
$c(3 \times 2)$	−15967.910
$c(4 \times 2)$	−23951.904
$c(5 \times 2)$	−31935.876
$c(6 \times 2)$	−39919.758
$c(3 \times 3)$	−31935.929
$c(4 \times 3)$	−47903.551
$c(4 \times 4)$	−71855.885

Table B.1: Reference total energies of the clean Pd(100) surface unit cells. Those values have been used according to the unit cell size of the calculated adsorbate configuration to calculate the corresponding CO binding energies which represented the quantity of interest for the parameterization of the cluster expansion.

B.1.2 On-site Binding Energy for One CO at a Pd(100) Bridge Site

As described in detail in subsection 8.3.2 before, the value for the on-site binding energy for the cluster expansion was extracted from a setup for a single CO molecule adsorbed on a bridge adsorption site in a $c(4 \times 4)$ surface unit cell.

Unit cell size	$E^{\text{total}}_{\text{bind}}$ in meV
$c(2 \times 2)$	−1946.2
$c(3 \times 3)$	−1954.8
$c(4 \times 4)$	−1957.3

Table B.2: Binding energy values for different unit cell sizes. For all further calculations the value for the largest unit cell size has been taken.

B.2 DFT Structure Database

B.2.1 Experimental Structures

c(3x2) 2 CO experimental structure

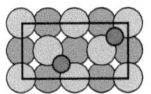

$c(3 \times 2)$ 2 CO experimental structure

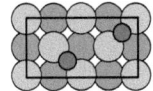

after relaxation
$E_{\text{bind}}^{\text{DFT}} = -3885.67\,\text{meV}$

c(4x2) 4 CO experimental structure

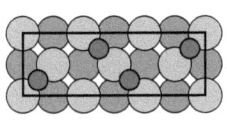

$c(4 \times 2)$ 4 CO experimental structure

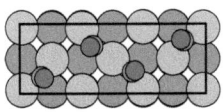

after relaxation
$E_{\text{bind}}^{\text{DFT}} = -7089.92\,\text{meV}$

c(5x2) 6 CO experimental structure

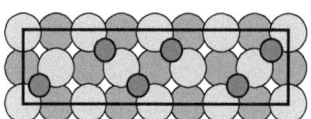

$c(5 \times 2)$ 6 CO experimental structure

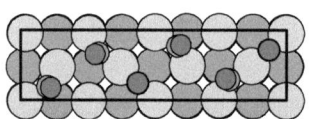

after relaxation
$E_{\text{bind}}^{\text{DFT}} = -10073.21\,\text{meV}$

B.2.2 c(2x2) structures

$c(2 \times 2)$ 1 CO

after relaxation
$E_{\text{bind}}^{\text{DFT}} = -1946.16\,\text{meV}$

B.2. DFT Structure Database

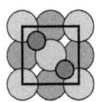

$c(2 \times 2)$ 2 CO A

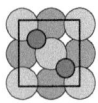

after relaxation
$E_{\text{bind}}^{\text{DFT}} = -2633.95$ meV

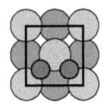

$c(2 \times 2)$ 2 CO B

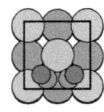

after relaxation
$E_{\text{bind}}^{\text{DFT}} = 780.98$ meV

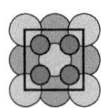

$c(2 \times 2)$ 4 CO, bridge double monolayer

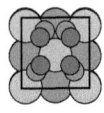

after relaxation
$E_{\text{bind}}^{\text{DFT}} = 9678.84$ meV

B.2.3 c(3x2) structure

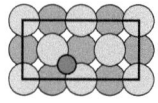

$c(3 \times 2)$ 1 CO

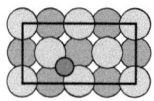

after relaxation
$E_{\text{bind}}^{\text{DFT}} = -2010.73 \text{ meV}$

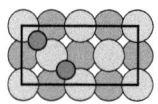

$c(3 \times 2)$ 2 CO A

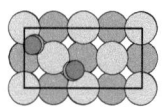

after relaxation
$E_{\text{bind}}^{\text{DFT}} = -3642.65 \text{ meV}$

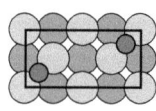

$c(3 \times 2)$ 2 CO B

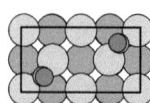

after relaxation
$E_{\text{bind}}^{\text{DFT}} = -3640.74 \text{ meV}$

B.2. DFT Structure Database

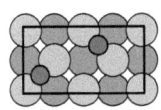

$c(3 \times 2)$ 2 CO C

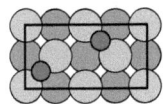

after relaxation
$E_{\text{bind}}^{\text{DFT}} = -3976.25$ meV

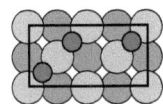

$c(3 \times 2)$ 3 CO A

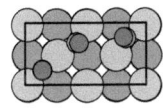

after relaxation
$E_{\text{bind}}^{\text{DFT}} = -5067.76$ meV

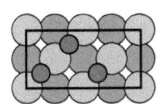

$c(3 \times 2)$ 3 CO B

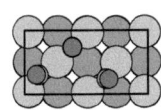

after relaxation
$E_{\text{bind}}^{\text{DFT}} = -5066.86$ meV

142   Appendix B. Adsorption Structure Database for the Cluster Expansion

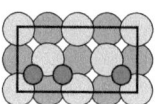

$c(3 \times 2)$ 3 CO C

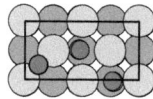

after relaxation
$E_{\text{bind}}^{\text{DFT}} = -4879.10\,\text{meV}$

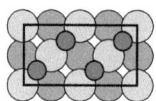

$c(3 \times 2)$ 4 CO A

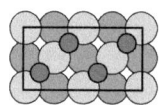

after relaxation
$E_{\text{bind}}^{\text{DFT}} = -5322.16\,\text{meV}$

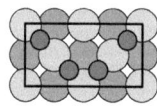

$c(3 \times 2)$ 4 CO B

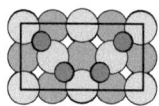

after relaxation
$E_{\text{bind}}^{\text{DFT}} = -3369.11\,\text{meV}$

B.2. DFT Structure Database

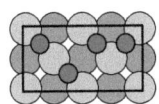

$c(3 \times 2)$ 4 CO C

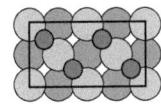

after relaxation
$E_{\text{bind}}^{\text{DFT}} = -5321.88$ meV

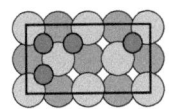

$c(3 \times 2)$ 4 CO D

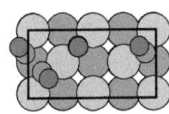

after relaxation
$E_{\text{bind}}^{\text{DFT}} = -4445.16$ meV

B.2.4 c(3x3) structures

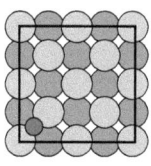

$c(3 \times 3)$ 1 CO

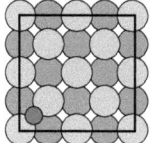

after relaxation
$E_{\text{bind}}^{\text{DFT}} = -1954.82\,\text{meV}$

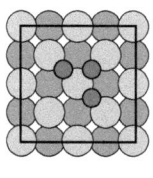

$c(3 \times 3)$ 3 CO A

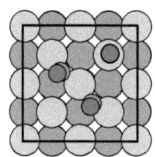

after relaxation
$E_{\text{bind}}^{\text{DFT}} = -5053.49\,\text{meV}$

B.2. DFT Structure Database

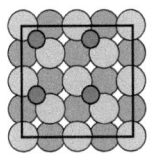

$c(3 \times 3)$ 4 CO A

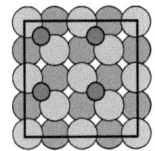

after relaxation
$E_{\text{bind}}^{\text{DFT}} = -7785.26\,\text{meV}$

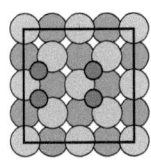

$c(3 \times 3)$ 4 CO B

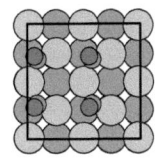

after relaxation
$E_{\text{bind}}^{\text{DFT}} = -7083.00\,\text{meV}$

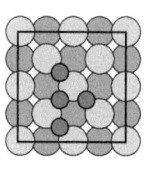
$c(3 \times 3)$ 4 CO C

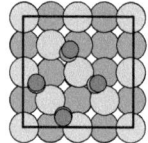
after relaxation
$E_{\text{bind}}^{\text{DFT}} = -7069.11 \text{ meV}$

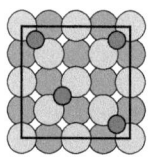
$c(3 \times 3)$ 4 CO D

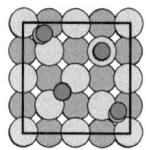
after relaxation
$E_{\text{bind}}^{\text{DFT}} = -6980.07 \text{ meV}$

B.2. DFT Structure Database

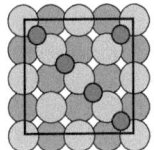

$c(3 \times 3)$ 5 CO A

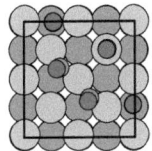

after relaxation
$E_{\text{bind}}^{\text{DFT}} = -8464.61$ meV

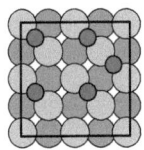

$c(3 \times 3)$ 5 CO B

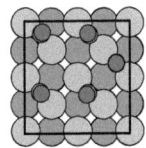

after relaxation
$E_{\text{bind}}^{\text{DFT}} = -8953.27$ meV

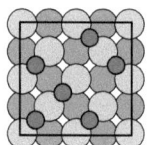

$c(3 \times 3)$ 6 CO A

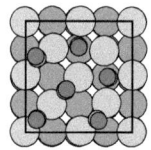

after relaxation
$E_{\text{bind}}^{\text{DFT}} = -9912.90 \text{ meV}$

B.2.5 c(4x2) structures

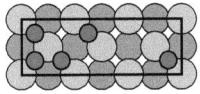

$c(4 \times 2)$ 5 CO A

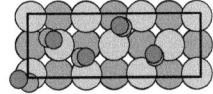

after relaxation
$E_{\text{bind}}^{\text{DFT}} = -6292.71\,\text{meV}$

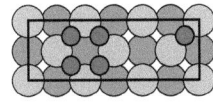

$c(4 \times 2)$ 5 CO B

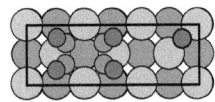

after relaxation
$E_{\text{bind}}^{\text{DFT}} = 1111.25\,\text{meV}$

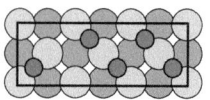

$c(4 \times 2)$ 5 CO C

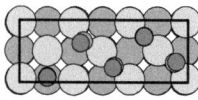

after relaxation
$E_{\text{bind}}^{\text{DFT}} = -7822.29\,\text{meV}$

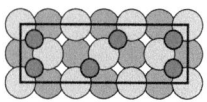

$c(4 \times 2)$ 6 CO A

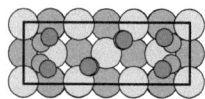

after relaxation
$E_{\text{bind}}^{\text{DFT}} = -358.28\,\text{meV}$

B.2.6 c(4x3) structures

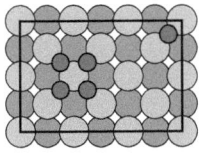

$c(4 \times 3)$ 5 CO A

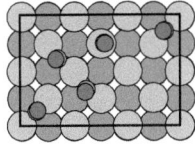

after relaxation
$E_{\text{bind}}^{\text{DFT}} = -9108.00$ meV

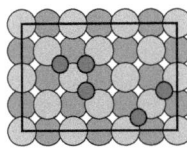

$c(4 \times 3)$ 5 CO B

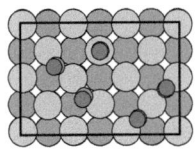

after relaxation
$E_{\text{bind}}^{\text{DFT}} = -9072.67$ meV

152 Appendix B. Adsorption Structure Database for the Cluster Expansion

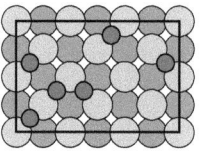

$c(4 \times 3)$ 6 CO A

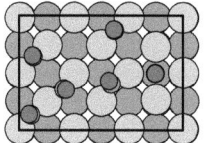

after relaxation
$E_{\text{bind}}^{\text{DFT}} = -11449.33$ meV

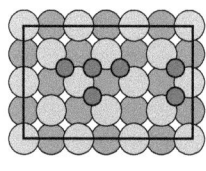

$c(4 \times 3)$ 6 CO B

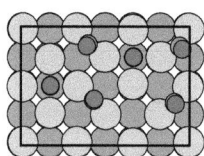

after relaxation
$E_{\text{bind}}^{\text{DFT}} = -11574.50$ meV

B.2. DFT Structure Database

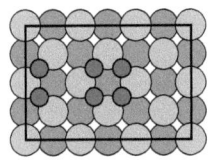

$c(4 \times 3)$ 6 CO C

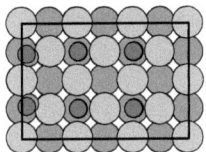

after relaxation
$E_{\text{bind}}^{\text{DFT}} = -11245.30\,\text{meV}$

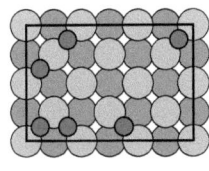

$c(4 \times 3)$ 6 CO D

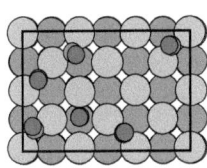

after relaxation
$E_{\text{bind}}^{\text{DFT}} = -11064.48\,\text{meV}$

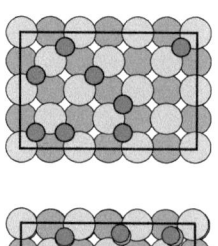

$c(4 \times 3)$ 8 CO A

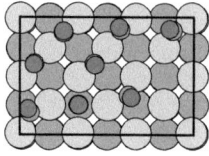

after relaxation
$E_{\text{bind}}^{\text{DFT}} = -13990.40\,\text{meV}$

B.2.7 The 10 additional c(4x3) adsorption structures

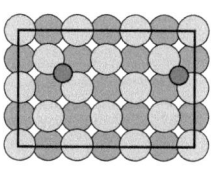

$c(4 \times 3)$ 2 CO

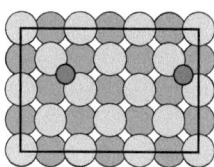

after relaxation
$E_{\text{bind}}^{\text{DFT}} = -4120.83\,\text{meV}$

B.2. DFT Structure Database

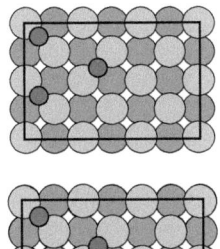

$c(4 \times 3)$ 3 CO A

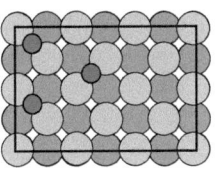

after relaxation
$E_{\text{bind}}^{\text{DFT}} = -6173.71 \text{ meV}$

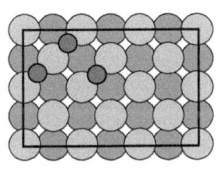

$c(4 \times 3)$ 3 CO B

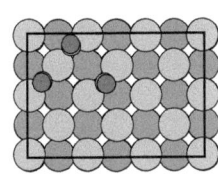

after relaxation
$E_{\text{bind}}^{\text{DFT}} = -5860.00 \text{ meV}$

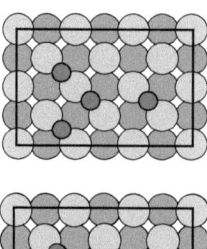 $c(4 \times 3)$ 4 CO A

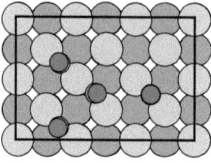 after relaxation
$E_{\text{bind}}^{\text{DFT}} = -7814.94 \, \text{meV}$

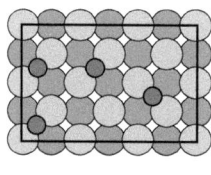 $c(4 \times 3)$ 4 CO B

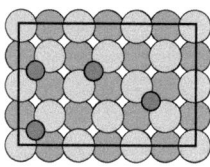 after relaxation
$E_{\text{bind}}^{\text{DFT}} = -8156.01 \, \text{meV}$

B.2. DFT Structure Database

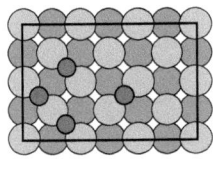

$c(4 \times 3)$ 4 CO C

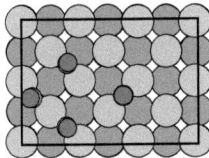

after relaxation
$E_{\text{bind}}^{\text{DFT}} = -7861.85\,\text{meV}$

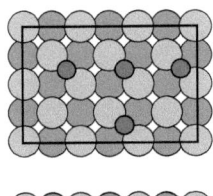

$c(4 \times 3)$ 4 CO D

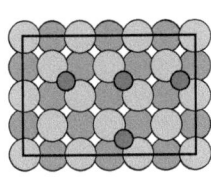

after relaxation
$E_{\text{bind}}^{\text{DFT}} = 8147.14\,\text{meV}$

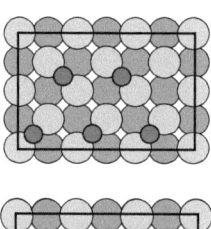

$c(4 \times 3)$ 5 CO C

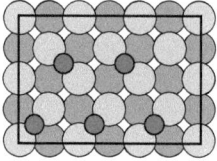

after relaxation
$E_{\text{bind}}^{\text{DFT}} = -10194.31\,\text{meV}$

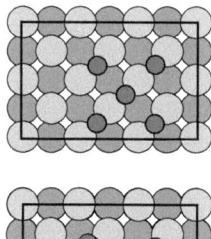

$c(4 \times 3)$ 5 CO D

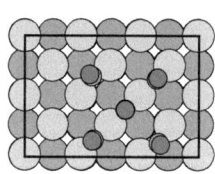

after relaxation
$E_{\text{bind}}^{\text{DFT}} = -9413.47\,\text{meV}$

B.2. DFT Structure Database

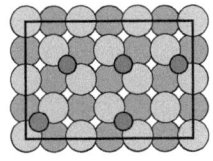

$c(4 \times 3)$ 5 CO E

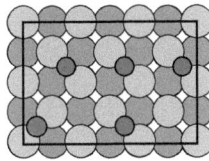

after relaxation
$E_{\text{bind}}^{\text{DFT}} = -10145.68\,\text{meV}$

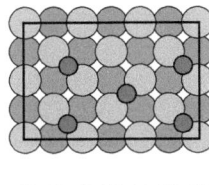

$c(4 \times 3)$ 5 CO F

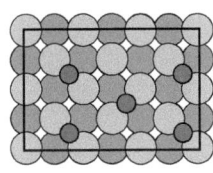

after relaxation
$E_{\text{bind}}^{\text{DFT}} = -10158.09\,\text{meV}$

B.2.8 c(4×4) structures

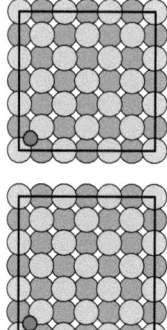

$c(4 \times 4)$ 1 CO

after relaxation
$E_{\text{bind}}^{\text{DFT}} = -1957.30$ meV

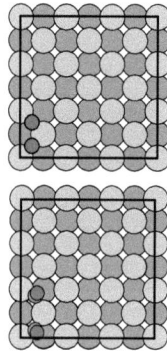

$c(4 \times 4)$ 2 CO

after relaxation
$E_{\text{bind}}^{\text{DFT}} = -3537.37$ meV

B.2. DFT Structure Database

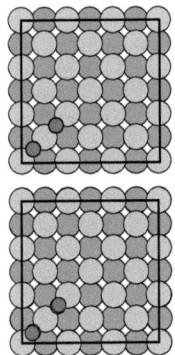

$c(4 \times 4)$ 2 CO A

after relaxation
$E_{\text{bind}}^{\text{DFT}} = -3688.17\,\text{meV}$

B.2.9 c(5×2) structures

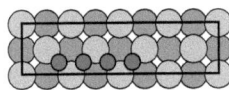

$c(5 \times 2)$ 4 CO A

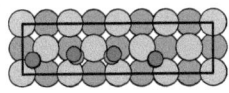

after relaxation
$E_{\text{bind}}^{\text{DFT}} = -7422.47\,\text{meV}$

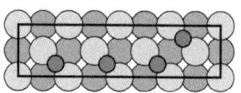

$c(5 \times 2)$ 4 CO B

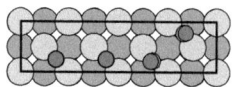

after relaxation
$E_{\text{bind}}^{\text{DFT}} = -7527.57$ meV

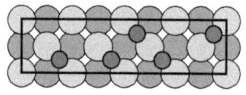

$c(5 \times 2)$ 5 CO A

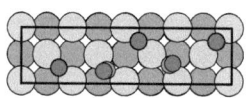

after relaxation
$E_{\text{bind}}^{\text{DFT}} = -9060.66$ meV

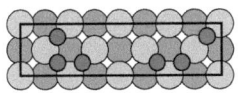

$c(5 \times 2)$ 6 CO A

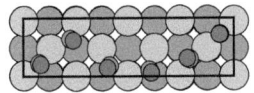

after relaxation
$E_{\text{bind}}^{\text{DFT}} = -9928.59\,\text{meV}$

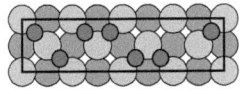

$c(5 \times 2)$ 7 CO A

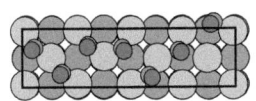

after relaxation
$E_{\text{bind}}^{\text{DFT}} = -10216.83\,\text{meV}$

164   Appendix B. Adsorption Structure Database for the Cluster Expansion

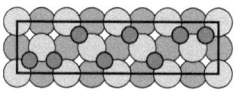 $c(5 \times 2)$ 8 CO A

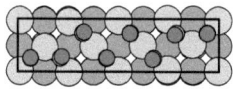 after relaxation
$E_{\text{bind}}^{\text{DFT}} = -8989.45\,\text{meV}$

B.2.10 c(6x2) structures

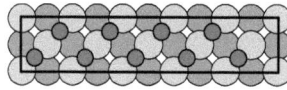 $c(6 \times 2)$ 9 CO

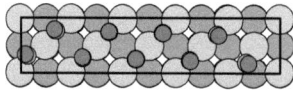 after relaxation
$E_{\text{bind}}^{\text{DFT}} = -13107.19\,\text{meV}$

Appendix C
Overview of Calculated Cu Island Structures

C.1 Structures Calculated on a DFT Level

In this section, the resulting geometries and energies of the different island structures calculated with density-functional theory are presented. In addition, the energy differences calculated with the embedded-atom method (EAM) are given in tabular form. In all cases structural relaxations started from «ideal» starting positions, where the Cu atoms either occupied the ideal hollow adsorption site of the underlying Ag(100) lattice, or were placed on the ideal bridge positions of the surface. DFT relaxations were performed using the BFGS algorithm as implemented in CASTEP[74]. All EAM relaxations were performed using the LAMMPS[188] molecular dynamics code and the implemented minimization procedures there (steepest decent or conjugate gradient). The convergence criteria were set to machine precision.

Calculation Parameters of DFT calculations

- Cut-off energy: 400 meV

- k-point sampling: $12 \times 12 \times 1$ grid for the $p(1 \times 1)$ cell, for larger unit cells the sampling grid was reduced to a $2 \times 2 \times 1$ grid

- Grid scale: 1.80

- Smearing width: 0.1 meV, Hermite Polynomials

- Force convergence criteria: $\Delta F \leq 3.5 \cdot 10^{-2}$ eV$\cdot$Å^{-1}

- Energy convergence criteria: $\Delta E \leq 1 \cdot 10^{-6}$ eV/atom

- Electronic minimization convergence criteria: $\Delta E \leq 1 \cdot 10^{-5}$ eV/atom

- XC-functional: PBE

C.2 DFT Geometries

C.2.1 3×3 islands

Final structure, starting from all-hollow

Final structure, starting from all-bridge

C.2.2 4×4 islands

Final structure, starting from all-hollow

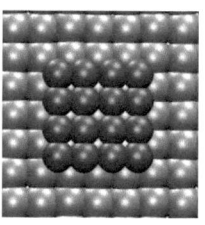

Final structure, starting from all-bridge

C.2.3 5×5 islands

Final structure, starting from all-hollow

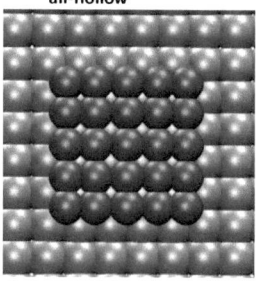

Final structure, starting from all-bridge

C.2.4 6 × 6 islands

Final structure, starting from all-hollow

Final structure, starting from all-bridge

C.2.5 7 × 7 islands

Final structure, starting from all-hollow

Final structure, starting from all-bridge

C.3 Comparison of Energy Differences for Different Island Sizes

Cu island size	DFT value	EAM value
	ΔE/Cu in eV	
4 × 4	+0.083	+0.117
5 × 5	+0.045	+0.082
6 × 6	+0.070	+0.007
7 × 7	+0.065	+0.066
9 × 9	--	+0.015
10 × 10	--	+0.001
monolayer	−0.434	−0.346

Table C.1: Stabilization energy per Cu atom for initially square Cu islands on Ag(100) after a DFT-BFGS geometry relaxation and after conjugate gradient EAM calculations.

Bibliography

[1] A. Zangwill, *Physics at Surfaces* (Cambridge University Press, UK, 1988).

[2] G. Ertl, *Interactions in Adsorbed Layers*, J. Vac. Sci. Technol. **14**, 435 (1977).

[3] N. Knorr, H. Brune, M. Epple, A. Hirstein, M. Schneider, and K. Kern, *Long-range Adsorbate Interactions Mediated by a Two-Dimensional Electron Gas*, Phys. Rev. B **65**, 115420 (2002).

[4] A. M. Bradshaw and M. Scheffler, *Lateral Interactions in Adsorbed Layers*, J. Vac. Sci. Technol. **16**, 447 (1979).

[5] J. Jones, *On the Determination of Molecular Fields. I. From the Variation of the Viscosity of a Gas with Temperature*, Proceedings of the Royal Society of London. Series A **106**, 441 (1924).

[6] J. Jones, *On the Determination of Molecular Fields. II. From the Equation of State of a Gas*, Proceedings of the Royal Society of London. Series A **106**, 463 (1924).

[7] M. Scheffler, *The Influence of Lateral Interactions on the Vibrational Spectrum of Adsorbed CO*, Surf. Sci. **81**, 562 (1979).

[8] J. Friedel, *Metallic Alloys*, Il Nuovo Cimento (1955-1965) **7**, 287 (1958).

[9] T. Kontorova and Y. Frenkel, *On the Theory of Plastic Deformation and Twinning I, II*, Zh. Eksp. & Teor. Fiz. **8**, 1349 (1938).

[10] Y. Frenkel and T. Kontorova, *On the Theory of Plastic Deformation and Twinning*, J. Phys. Acad. Sci. USSR **1**, 149 (1939).

[11] O. Braun and Y. Kivshar, *Nonlinear Dynamics of the Frenkel-Kontorova Model*, Phys. Rep. **306**, 1 (1998).

[12] M. Mansfield and R. Needs, *Application of the Frenkel-Kontorova Model to Surface Reconstructions*, J. Phys.: Condens. Matt. **2**, 2361 (1990).

[13] I. Lyuksyutov, A. Naumovets, and V. Pokrovsky, *Two-Dimensional Crystals* (Naukova Dumka, Kiev (English Translation: Academic Press, Boston, 1992), 1988).

[14] J. C. Hamilton and S. M. Foiles, *Misfit Dislocation Structure for Close-Packed Metal-Metal Interfaces*, Phys. Rev. Lett. **75**, 882 (1995).

[15] J. C. Hamilton, *Overlayer Strain Relief on Surfaces with Square Symmetry: Phase Diagram for a 2D Frenkel-Kontorova Model*, Phys. Rev. Lett. **88**, 126101 (2002).

[16] J. C. Hamilton, *Magic Size Effects for Heteroepitaxial Island Diffusion*, Phys. Rev. Lett. **77**, 885 (1996).

[17] G. Comsa, K. Kern, and B. Poelsema, in E. Hulpke, ed., *Springer Series in Surface Science* (Springer Verlag, Berlin, Germany, 1992), vol. 27, chap. 10, pp. 243–264.

[18] Y. Zhang, *First-Principles Statistical Mechanics Approach to Step Decoration at Solid Surfaces*, Ph.D. thesis, Freie Universität Berlin (2008).

[19] C. Zaum, *Schnelle Rastertunnelmikroskopie: Entwicklung einer Echtzeitsteuerung und Dynamik heteroepitaktischer Cu-Inseln auf Ag(100)*, Master's thesis, Leibniz Universität Hannover (2007).

[20] C. Herring, *Some Theorems on the Free Energy of Crystal Surfaces*, Phys. Rev. **82**, 87 (1951).

[21] W. Tyson and W. Miller, *Surface Free Energies of Solid Metals: Estimation from Liquid Surface Tension Measurements*, Surf. Sci. **62**, 267 (1977).

[22] A. Szabo and N. Ostlund, *Modern Quantum Chemistry* (Dover Publications, New York, 1989).

[23] W. Koch and M. Holthausen, *A Chemist's Guide to Density Functional Theory* (Wiley, VCH, 1999).

[24] E. Gross and R. Dreizler, *Density Functional Theory* (Springer Verlag, Berlin, Germany, 1990).

[25] R. Parr and W. Yang, *Density-Functional Theory of Atoms and Molecules* (Oxford University Press, 1989).

[26] E. Schrödinger, *Quantisierung als Eigenwertproblem*, Ann. Phys. **384**, 361 (1926).

[27] M. Born and R. Oppenheimer, *Zur Quantentheorie der Molekeln*, Ann. Phys. **389**, 457 (1927).

[28] W. Pauli, *Über den Zusammenhang des Abschlusses der Elektronengruppen im Atom mit der Komplexstruktur der Spektren*, Zeitschrift für Physik A **31**, 765 (1925).

[29] W. Pauli, *The Connection Between Spin and Statistics*, Phys. Rev. **58**, 716 (1940).

[30] W. Ritz, *Über eine neue Methode zur Lösung gewisser Variationsprobleme der mathematischen Physik.*, Journal für die Reine und Angewandte Mathematik **135**, 1 (1909).

[31] J. K. L. MacDonald, *Successive Approximations by the Rayleigh-Ritz Variation Method*, Phys. Rev. **43**, 830 (1933).

[32] D. R. Hartree, *The Wave Mechanics of an Atom with a Non-Coulomb Central Field. Part I. Theory and Methods*, Math. Proc. Cam. Phil. Soc. **24**, 89 (1928).

[33] D. R. Hartree, *The Wave Mechanics of an Atom with a Non-Coulomb Central Field. Part II. Some Results and Discussion*, Math. Proc. Cam. Phil. Soc. **24**, 111 (1928).

[34] F. J. Dyson and A. Lenard, *Stability of Matter. I*, J. Math. Phys. **8**, 423 (1967).

[35] E. H. Lieb, *The Stability of Matter*, Rev. Mod. Phys. **48**, 553 (1976).

[36] J. C. Slater, *The Theory of Complex Spectra*, Phys. Rev. **34**, 1293 (1929).

[37] V. Fock, *Näherungsmethode zur Lösung des quantenmechanischen Mehrkörperproblems*, Zeitschrift für Physik A **61**, 126 (1930).

[38] F. Jensen, *Introduction to Computational Chemistry* (Wiley, VCH, 1999).

[39] C. Møller and M. Plesset, *Note on an Approximation Treatment for Many-Electron Systems*, Phys. Rev. **46**, 0618 (1934).

[40] T. Dunning, *A Road Map for the Calculation of Molecular Binding Energies*, J. Phys. Chem. A **104**, 9062 (2000).

[41] L. H. Thomas, *The Calculation of Atomic Fields*, Math. Proc. Cam. Phil. Soc. **23**, 542 (1926).

[42] E. Fermi, *Eine statistische Methode zur Bestimmung einiger Eigenschaften des Atoms und ihre Anwendung auf die Theorie des periodischen Systems der Elemente*, Zeitschrift für Physik A **48**, 73 (1928).

[43] R. O. Jones and O. Gunnarsson, *The Density Functional Formalism, its Applications and Prospects*, Rev. Mod. Phys. **61**, 689 (1989).

[44] E. Teller, *On the Stability of Molecules in the Thomas-Fermi Theory*, Rev. Mod. Phys. **34**, 627 (1962).

[45] N. L. Baláza, *Formation of Stable Molecules within the Statistical Theory of Atoms*, Phys. Rev. **156**, 42 (1967).

[46] E. H. Lieb and B. Simon, *Thomas-Fermi Theory Revisited*, Phys. Rev. Lett. **31**, 681 (1973).

[47] P. A. M. Dirac, *Note on Exchange Phenomena in the Thomas Atom*, Math. Proc. Cam. Phil. Soc. **26**, 376 (1930).

[48] E. Wigner and F. Seitz, *On the Constitution of Metallic Sodium. II*, Phys. Rev. **46**, 509 (1934).

[49] P. Hohenberg and W. Kohn, *Inhomogeneous Electron Gas*, Phys. Rev. **136**, B864 (1964).

[50] W. Kohn and L. Sham, *Self-Consistent Equations Including Exchange and Correlation Effects*, Phys. Rev. **140**, A1133 (1965).

[51] J. Chayes, L. Chayes, and M. Ruskai, *Density Functional Approach to Quantum Lattice Systems*, Journal of Statistical Physics **38**, 497 (1985).

[52] P. Noziéres and D. Pines, *Correlation Energy of a Free Electron Gas*, Phys. Rev. **111**, 442 (1958).

[53] W. J. Carr, *Energy, Specific Heat, and Magnetic Properties of the Low-Density Electron Gas*, Phys. Rev. **122**, 1437 (1961).

[54] M. Gell-Mann and K. Brueckner, *Correlation Energy of an Electron Gas at High Density*, Phys. Rev. **106**, 364 (1957).

[55] W. J. Carr and A. A. Maradudin, *Ground-State Energy of a High-Density Electron Gas*, Phys. Rev. **133**, A371 (1964).

[56] D. M. Ceperly and B. J. Alder, *Ground State of the Electron Gas by a Stochastic Method*, Phys. Rev. Lett. **45**, 566 (1980).

[57] S. H. Vosko, L. Wilk, and M. Nusair, *Accurate Spin-Dependent Electron Liquid Correlation Energies for Local Spin Density Calculations: A Critical Analysis*, Canadian Journal of Physics (1980).

[58] J. P. Perdew and A. Zunger, *Self-Interaction Correction to Density-Functional Approximations for Many-Electron Systems*, Phys. Rev. B **23**, 5048 (1981).

[59] J. P. Perdew and Y. Wang, *Accurate and Simple Analytic Representation of the Electron-Gas Correlation Energy*, Phys. Rev. B **45**, 13244 (1992).

[60] J. P. Perdew, K. Burke, and M. Ernzerhof, *Generalized Gradient Approximation Made Simple*, Phys. Rev. Lett. **77**, 3865 (1996).

[61] J. P. Perdew, K. Burke, and M. Ernzerhof, *Errata: Generalized Gradient Approximation Made Simple*, Phys. Rev. Lett. **78**, 1396 (1997).

[62] B. Meyer, *The Pseudopotential Plane Wave Approach*, NIC Series **31**, 71 (2006).

[63] F. Bloch, *Über die Quantenmechanik der Elektronen in Kristallgittern*, Zeitschrift für Physik A **52**, 555 (1928).

[64] D. J. Chadi and M. Cohen, *Special Points in the Brillouin Zone*, Phys. Rev. B **8**, 5747 (1973).

Bibliography

[65] H. J. Monkhorst and J. D. Pack, *Special points for Brillouin-zone Integrations*, Phys. Rev. B **13**, 5188 (1976).

[66] G. Gilat and N. R. Bharatiya, *Tetrahedron Method of Zone Integration: Inclusion of Matrix Elements*, Phys. Rev. B **12**, 3479 (1975).

[67] P. E. Blöchl, O. Jepsen, and O. K. Andersen, *Improved Tetrahedron Method for Brillouin-zone Integrations*, Phys. Rev. B **49**, 16223 (1994).

[68] J. Moreno and J. Soler, *Optimal Meshes for Integrals in Real- and Reciprocal-Space Unit Cells*, Phys. Rev. B **45**, 13891 (1992).

[69] C.-L. Fu and K.-M. Ho, *First-Principles Calculation of the Equilibrium Ground-State Properties of Transition Metals: Applications to Nb and Mo*, Phys. Rev. B **28**, 5480 (1983).

[70] M. J. Gillian, *Calculation of the Vacancy Formation Energy in Aluminium*, J. Phys.: Condens. Matt. **1**, 689 (1989).

[71] M. Methfessel and S. H. Paxton, *High-Precision Sampling for Brillouin-Zone Integration in Metals*, Phys. Rev. B **40**, 3616 (1989).

[72] K. Laasonen, A. Pasquarello, R. Car, C. Lee, and D. Vanderbilt, *Car-Parrinello Molecular Dynamics with Vanderbilt Ultrasoft Pseudopotentials*, Phys. Rev. B **47**, 10142 (1993).

[73] D. Vanderbilt, *Soft Self-Consistent Pseudopotentials in a Generalized Eigenvalue Formalism*, Phys. Rev. B **41**, 7892 (1990).

[74] S. J. Clark, M. Segall, C. Pickard, and P. J. Hasnip, *First-Principles Methods Using CASTEP*, Z. Kristallogr. (2005).

[75] J. Bardeen, *Tunnelling from a Many-Particle Point of View*, Phys. Rev. Lett. **6**, 57 (1961).

[76] J. Tersoff and D. R. Hamann, *Theory and Application for the Scanning Tunneling Microscope*, Phys. Rev. Lett. **50**, 1998 (1983).

[77] J. Tersoff and D. R. Hamann, *Theory of the Scanning Tunneling Microscope*, Phys. Rev. B **31**, 805 (1985).

[78] R. Wiesendanger, *Scanning Probe Microscopy and Spectroscopy* (Cambridge, New York, 1994).

[79] D. Chandler, *Introduction to Modern Statistical Mechanics* (Oxford University Press, 1987).

[80] J. Sanchez, F. Ducastelle, and D. Gratias, *Generalized Cluster Description of Multicomponent Systems*, Physica A **128**, 334 (1984).

[81] R. Kikuchi, *A Theory of Cooperative Phenomena*, Phys. Rev. **81**, 988 (1951).

[82] A. Zunger, in P. Turchy and A. Gonis, eds., *Statics and Dynamics of Alloy Phase Transformations* (NATO ASI Series, Plenum Press, New York, 1994).

[83] S. Müller, M. Stöhr, and O. Wieckhorst, *Structure and stability of binary alloy surfaces: Segregation, relaxation, and ordering from first-principles calculations*, Applied Physics A: Materials Science & Processing **82**, 415 (2006).

[84] C. Stampfl, H. J. Kreuzer, S. H. Paxton, H. Pfnür, and M. Scheffler, *First-Principles Theory of Surface Thermodynamics and Kinetics*, Phys. Rev. Lett. **83**, 2993 (1999).

[85] R. Drautz, R. Singer, and M. Fähnle, *Cluster expansion technique: An efficient tool to search for ground-state configurations of adatoms on plane surfaces*, Phys. Rev. B **67**, 035418 (2003).

[86] S. Müller, *Bulk and Surface Ordering Phenomena in Binary Metal Alloys*, J. Phys.: Condens. Matt. **15**, R1429 (2003).

[87] K. Reuter, C. Stampfl, and M. Scheffler, in S. Yip, ed., *Handbook of Materials Modeling* (Springer Dordrecht, Netherlands, 2005), pp. 149–234.

[88] S. D. Miller and J. R. Kitchin, *Uncertainty and Figure Selection for DFT Based Cluster Expansions for Oxygen Adsorption on Au and Pt (111) Surfaces*, Mol. Sim. **35**, 920 (2009).

[89] D. Lerch, O. Wieckhorst, G. L. Hart, R. W. Forcade, and S. Müller, *UNCLE: a code for constructing cluster expansions for arbitrary lattices with minimal user-input*, Modelling Simul. Mater. Sci. Eng. **17**, 055003 (2009).

[90] J. Connolly and A. Williams, *Density-Functional Theory Applied to Phase Transformations in Transition-Metal Alloys*, Phys. Rev. B **27**, 5169 (1983).

[91] V. Blum and A. Zunger, *Mixed-Basis Cluster Expansion for Thermodynamics of bcc Alloys*, Phys. Rev. B **70**, 155108 (2004).

[92] Q.-S. Xu, Y.-Z. Liang, and Y.-P. Du, *Monte Carlo Cross-Validation for Selecting a Model and Estimating the Prediction Error in Multivariate Calibration*, J. Chemometrics **18**, 112 (2004).

[93] J. Shao, *Linear Model Selection by Cross-Validation*, J. Amer. Statist. Assoc. **88**, 486 (1993).

[94] R. G. Caflisch and A. N. Berker, *Oxygen Chemisorbed on Ni(100): A Renormalization-Group Study of the Global Phase Diagram*, Phys. Rev. B **29**, 1279 (1984).

[95] A. van de Walle and G. Ceder, *Automating First-Principles Phase Diagram Calculations*, Journal of Phase Equilibria **23**, 348 (2002).

[96] G. D. Garbulsky and G. Ceder, *Linear-Programming Method for Obtaining Effective Cluster Interactions in Alloys From Total-Energy Calculations: Application to the fcc Pd-V System*, Phys. Rev. B **51**, 67 (1995).

[97] K. Reuter and M. Scheffler, *Composition, Structure and Stability of $RuO_2(110)$ as a Function of Oxygen Pressure*, Phys. Rev. B **65**, 035406 (2001).

[98] M. Ernzerhof and G. E. Scuseria, *Assessment of the Perdew–Burke–Ernzerhof Exchange-Correlation Functional*, J. Chem. Phys. **110**, 5029 (1999).

[99] Y. Zhang, V. Blum, and K. Reuter, *Accuracy of First-Principles Lateral Interactions: Oxygen at Pd(100)*, Phys. Rev. B **75**, 235406 (2007).

[100] M. S. Daw and M. I. Baskes, *Semiempirical, Quantum Mechanical Calculation of Hydrogen Embrittlement in Metals*, Phys. Rev. Lett. **50**, 1285 (1983).

[101] M. S. Daw and M. I. Baskes, *Embedded-Atom Method: Derivation and Application to Impurities, Surfaces, and other Defects in Metals*, Phys. Rev. B **29**, 6443 (1984).

[102] S. M. Foiles, M. I. Baskes, and M. S. Daw, *Embedded-Atom-Method Functions for the fcc Metals Cu, Ag, Au, Ni, Pd, Pt, and their Alloys*, Phys. Rev. B **33**, 7983 (1986).

[103] S. M. Foiles, M. I. Baskes, and M. S. Daw, *Erratum: Embedded-Atom-Method Functions for the fcc Metals Cu, Ag, Au, Ni, Pd, Pt, and their Alloys*, Phys. Rev. B **37**, 10378 (1988).

[104] M. S. Daw, *Model of Metallic Cohesion: The Embedded-Atom-Method*, Phys. Rev. B **39**, 7441 (1989).

[105] Y. Mishin, in S. Yip, ed., *Handbook of Materials Modeling* (Springer Dordrecht, Netherlands, 2005), vol. 1, pp. 459–478.

[106] M. S. Daw, S. M. Foiles, and M. I. Baskes, *Embedded-Atom-Method – A Review of Theory and Applications*, Mat. Sci. Rep. **9**, 251 (1993).

[107] P. Williams, Y. Mishin, and J. C. Hamilton, *An Embedded-Atom Potential for the Cu-Ag System*, Modelling Simul. Mater. Sci. Eng. (2006).

[108] G. Kresse and J. Furthmüller, *Efficient Iterative Schemes for ab initio Total-Energy Calculations Using a Plane-Wave Basis Set*, Phys. Rev. B **54**, 11169 (1996).

[109] Y. Mishin, M. J. Mehl, D. A. Papaconstantopoulos, A. F. Voter, and J. D. Kress, *Structural Stability and Lattice Defects in Copper: Ab initio, Tight-Binding, and Embedded-Atom Calculations*, Phys. Rev. B **63**, 224106 (2001).

[110] K. Terakura, T. Oguchi, T. Mohri, and K. Watanabe, *Electronic Theory of the Alloy Phase Stability of Cu-Ag, Cu-Au, and Ag-Au Systems*, Phys. Rev. B **35**, 2169 (1987).

[111] R. Najafabadi, D. Srolovitz, E. Ma, and M. Atzmon, *Thermodynamic Properties of Metastable Ag-Cu Alloys*, J. Appl. Phys. **74**, 3144 (1993).

[112] C. Becker, *Interatomic potentials repository project*, http://www.ctcms.nist.gov/potentials/.

[113] R. Fletcher and C. M. Reeves, *Function Minimization by Conjugate Gradients*, The Computer Journal **7**, 149 (1964).

[114] W. K. Press, S. A. Teukolsky, W. T. Vetterlin, and B. T. Flannery, *Numerical Recipes* (Cambridge University Press, UK, 2007), 3rd ed.

[115] S. Kirkpatrick, C. Gelatt, and M. Vecchi, *Optimization by Simulated Annealing*, Science **220**, 671 (1983).

[116] N. Metropolis, A. W. Rosenbluth, M. N. Rosenbluth, and A. H. Teller, *Equation of State Calculations by Fast Computing Machines*, J. Chem. Phys. **21**, 1087 (1953).

[117] Y. Xiang, D. Sun, W. Fan, and X. Gong, *Generalized Simulated Annealing Algorithm and its Application to the Thomson Model*, Phys. Lett. A **233**, 216 (1997).

[118] H. Szu and R. Hartley, *Fast Simulated Annealing*, Phys. Lett. A **122**, 157 (1987).

[119] L. Ingber, *Adaptive Simulated Annealing (ASA): Lessons Learned*, Control and Cybernetics **25**, 33 (1996).

[120] C. Pickard and R. Needs, *High-Pressure Phases of Silane*, Phys. Rev. Lett. **97**, 045504 (2006).

[121] A. F. Voter, F. Montalenti, and T. Germann, *Extending the Time Scale in a Atomistic Simulation of Materials*, Annu. Rev. Mat. Res. **32**, 321 (2002).

[122] M. Nic, J. Jirat, and B. K. et al., *Iupac gold book*, http://goldbook.iupac.org/about.html.

[123] D. Sheppard, R. Terrell, and G. Henkelman, *Optimization Methods for Finding Minimum Energy Paths*, J. Chem. Phys. **128**, 134106 (2008).

[124] R. Elber and M. Karplus, *A Method for Determining Reaction Paths in Large Molecules: Application to Myoglobin*, Chem. Phys. Lett. **139**, 375 (1987).

[125] L. Pratt, *A Statistical Method for Identifying Transition States in High Dimensional Problems*, J. Chem. Phys. **85**, 5045 (1986).

[126] H. Jónsson and G. Henkelman, *Improved Tangent Estimate in the Nudged Elastic Band Method for Finding Minimum Energy Paths and Saddle Points*, J. Chem. Phys. **113**, 9978 (2000).

[127] H. Eyring, *The Activated Complex in Chemical Reactions*, J. Chem. Phys. **3**, 107 (1935).

[128] E. Wigner, *The Transition State Method*, Trans. Faraday Soc. **34**, 29 (1938).

[129] J. C. Keck, in I. Prigogine, ed., *Advance in Chemical Physics* (Wiley, New York, 1967), vol. 13, chap. 5, pp. 85–121.

[130] A. F. Voter and J. Doll, *Dynamical Corrections to Transition State Theory for Multistate Systems: Surface Self-diffusion in the Rare-Event Regime*, J. Chem. Phys. **82**, 80 (1985).

[131] A. F. Voter and J. Doll, *Transition State Theory Description of Surface Self-Diffusion: Comparison with Classical Trajectory Results*, J. Chem. Phys. **80**, 5832 (1984).

[132] G. H. Vineyard, *Frequency Factors and Isotope Effects in Solid State Rate Processes*, J. Phys. Chem. Solids **3**, 121 (1957).

[133] C. Wert and C. Zener, *Interstitial Atomic Diffusion Coefficients*, Phys. Rev. **76**, 1169 (1949).

[134] D. Frenkel and B. Smit, *Understanding Molecular Simulation: From Algorithms to Applications* (Elsevier, 2002).

[135] A. F. Voter, *Parallel Replica Method for Dynamics of Infrequent Events*, Phys. Rev. B **57**, R13985 (1998).

[136] J. C. Tracy and P. W. Palmberg, *Simple Technique For Binding Energy Determinations: CO on Pd (100)*, Surf. Sci. **14**, 274 (1969).

[137] A. M. Bradshaw and F. M. Hoffmann, *The Chemisorption of Carbon Monoxide on Palladium Single Crystal Surfaces: IR Spectroscopic Evidence for Localised Site Adsorption*, Surf. Sci. **72**, 513 (1978).

[138] R. Behm, K. Christmann, G. Ertl, M. V. Hove, P. Thiel, and W. H. Weinberg, *The Structure of CO Adsorbed on Pd(100): A LEED and HREELS Analysis*, Surf. Sci. **88**, L59 (1979).

[139] R. Behm, K. Christmann, G. Ertl, and M. V. Hove, *Adsorption of CO on Pd (100)*, J. Chem. Phys. **73**, 2984 (1980).

[140] P. Uvdal, P.-A. Karlsson, C. Nyberg, S. Andersson, and N. V. Richardson, *On the Structure of Dense CO Overlayers*, Surf. Sci. **202**, 167 (1988).

[141] F. M. Hoffmann, *Infrared Reflection-Absorption Spectroscopy of Adsorbed Molecules*, Surf. Sci. Rep. **3**, 107 (1983).

[142] K. Yoshioka, F. Kitamura, M. Takeda, M. Takahashi, and M. Ito, *Infrared Reflection Absorption Spectra of Carbon Monoxide Adsorbed on Single Crystal Electrodes, Pd(111) and Pd(100)*, Surf. Sci. **227**, 90 (1990).

[143] J. N. Andersen, M. Qvarford, R. Nyholm, S. L. Sorensen, and C. Wigren, *Surface Core-Level Shifts as a Probe of the Local Overlayer Structure: CO on Pd(100)*, Phys. Rev. Lett. **67**, 2822 (1991).

[144] W. Berndt and A. M. Bradshaw, *Domain-Wall Superlattices in High Coverage CO Adlayers on Pd(100)*, Surf. Sci. Lett. **279**, L165 (1992).

[145] M. Tushaus, W. Berndt, H. Conrad, A. M. Bradshaw, and B. N. J. Persson, *Understanding the Structure of High Coverage CO Adlayers*, Appl. Phys. A **51**, 91 (1990).

[146] B. N. J. Persson, M. Tüshaus, and A. M. Bradshaw, *On the Nature of Dense CO Adlayers*, J. Chem. Phys. **92**, 5034 (1990).

[147] J. Szanyi, W. K. Kuhn, and D. W. Goodman, *CO Adsorption on Pd(111) and Pd(100): Low and High Pressure Correlations*, J Vac Sci Technol A **11**, 1969 (1993).

[148] Y. Yeo, L. Vattuone, and D. A. King, *Calorimetric Investigation of NO and CO Adsorption on Pd(100) and the Influence of Preadsorbed Carbon*, J. Chem. Phys. **106**, 1990 (1997).

[149] R. Schuster, I. Robinson, K. Kuhnke, S. Ferrer, J. Alvarez, and K. Kern, *Pokrovsky-Talapov commensurate-incommensurate transition in the CO/Pd(100) system*, Phys. Rev. B **54**, 17097 (1996).

[150] T. Mitsui, M. Rose, E. Fomin, and D. Ogletree, *Diffusion and Pair Interactions of CO Molecules on Pd (111)*, Phys. Rev. Lett. **94**, 036101 (2005).

[151] A. Eichler and J. Hafner, *Adsorption of CO on Pd (100): Steering into Less Favored Adsorption Sites*, Phys. Rev. B **57**, 10110 (1998).

[152] F. Delbecq and P. Sautet, *Density Functional Periodic Study of CO adsorption on the $Pd_3Mn(100)$ Alloy Surface: Comparison with Pd(100)*, Phys. Rev. B **59**, 5142 (1999).

[153] M. Gajdoš, A. Eichler, and J. Hafner, *CO Adsorption on Close-Packed Transition and Noble Metal Surfaces: Trends from ab initio Calculations*, J. Phys.: Condens. Matt. **16**, 1141 (2004).

[154] P. J. Feibelman, B. Hammer, J. K. Nørskov, M. Scheffler, R. Stumpf, R. Watwe, and J. Dumesic, *The CO/Pt(111) Puzzle*, J. Phys. Chem. B **105**, 4018 (2001).

[155] R. A. Olsen, H. T. Philipsen, and E. J. Baerends, *CO on Pt(111): A Puzzle Revisited*, J. Chem. Phys. **119**, 4522 (2003).

[156] D. L. S. Nieskens, D. Curulla-Ferré, and J. W. Niemantsverdriet, *Atom–Molecule Interactions on Transition Metal Surfaces : A DFT Study of CO and Several Atoms on Rh(100), Pd(100) and Ir(100)*, ChemPhysChem **7**, 1075 (2006).

[157] G. Blyholder, *Molecular Orbital View of Chemisorbed Carbon Monoxide*, J. Phys. Chem. **68**, 2772 (1964).

[158] G. Blyholder and M. C. Allen, *Infrared Spectra and Molecular Orbital Model for Carbon Monoxide Adsorbed on Metals*, J. Am. Chem. Soc. **91**, 3158 (1969).

[159] D.-J. Liu and J. Evans, *Atomistic Lattice-Gas Modeling of CO Oxidation on Pd (100): Temperature-Programed Spectroscopy and Steady-State Behavior*, J. Chem. Phys. **124**, 154705 (2006).

[160] D.-J. Liu and J. Evans, *Chemical Diffusion of CO in Mixed CO+O Adlayers and Reaction-Front Propagation in CO Oxidation on Pd(100)*, J. Chem. Phys. **125**, 054709 (2006).

[161] D.-J. Liu and J. Evans, *From Atomic Scale Reactant Ordering to Mesoscale Reaction Front Propagation: CO Oxidation on Pd(100)*, Phys. Rev. B **70**, 193408 (2004).

[162] D.-J. Liu, *Lattice-Gas Modeling of CO Adlayers on Pd(100)*, J. Chem. Phys. **121**, 4352 (2004).

[163] C. G. M. Hermse, A. Jansen, A. P. V. Bavel, J. J. Lukkien, and R. A. V. Santen, *On the Nature of Dense CO Adlayers on fcc(100) Surfaces: a Kinetic Monte Carlo Study*, Phys. Chem. Chem. Phys. **12**, 461 (2010).

[164] K. Kern, P. Zeppenfeld, R. David, and G. Comsa, *Structure and Dynamics of Rare-Gas Layers on Pt(111)*, Journal of Electron Spectroscopy and Related Phenomena **44**, 215 (1987).

[165] A. Tkatchenko and M. Scheffler, *Accurate Molecular Van Der Waals Interactions from Ground-State Electron Density and Free-Atom Reference Data*, Phys. Rev. Lett. **102**, 073005 (2009).

[166] E. M. McNellis, *First-Principles Modeling of Molecular Switches at Surfaces*, Ph.D. thesis, Freie Universität Berlin (2009).

[167] S. H. Wei, A. A. Mbaye, L. G. Ferreira, and A. Zunger, *First-Principles Calculations of the Phase Diagrams of Noble Metals: Cu-Au, Cu-Ag and Ag-Au*, Phys. Rev. B **36**, 4163 (1987).

[168] K. Terakura, T. Oguchi, T. Mohri, and K. Watanabe, *Electronic theory of the alloy phase stability of Cu-Ag, Cu-Au, and Ag-Au systems*, Phys. Rev. B (1987).

[169] K. Uenishi, K. Kobayashi, K. Ishihara, and P. Shingu, *Formation of a Super-Saturated Solid Solution in the Ag—Cu System by Mechanical Alloying*, Materials Science and Engineering: A **134**, 1342 (1991).

[170] M. Pfeifer, O. Robach, B. Ocko, and I. Robinson, *Thickness-Related Instability of Cu Thin Films on Ag(100)*, Physica B **357**, 152 (2005).

[171] M. Dietterle, T. Will, and D. Kolb, *The Initial Stages of Cu Electrodeposition on Ag(100): An in situ STM Study*, Surf. Sci. **396**, 189 (1998).

[172] J. Stevens and R. Hwang, *Strain Stabilized Alloying of Immiscible Metals in Thin Films*, Phys. Rev. Lett. **74**, 2078 (1995).

[173] S. Piccinin, C. Stampfl, and M. Scheffler, *First-Principles Investigation of Ag-Cu Alloy Surfaces in an Oxidizing Environment*, Phys. Rev. B **77**, 075426 (2008).

[174] K. Morgenstern and K. Rieder, *Long-range Interaction of Copper Adatoms and Copper Dimers on Ag (111)*, N. J. Phys. **7**, 139 (2005).

[175] K. Morgenstern, K. Braun, and K. Rieder, *Direct Imaging of Cu Dimer Formation, Motion, and Interaction with Cu Atoms on Ag (111)*, Phys. Rev. Lett. **93**, 056102 (2004).

[176] J. Tersoff, *Surface-Confined Alloy Formation in Immiscible Systems*, Phys. Rev. Lett. **74**, 434 (1995).

[177] *Webelements periodic table of the elements*, http://www.webelements.com/.

[178] A. Einstein, *Über die von der molekularkinetischen Theorie der Wärme geforderte Bewegung von in ruhenden Flüssigkeiten suspendierten Teilchen, 1905*, Annalen der Physik **14**, 182 (2005).

[179] M. von Smoluchowski, *Zur kinetischen Theorie der Brownschen Molekularbewegung und der Suspensionen*, Annalen der Physik **326**, 756 (1906).

[180] A. F. Voter, *Classically Exact Overlayer Dynamics: Diffusion of Rhodium Clusters on Rh (100)*, Phys. Rev. B **34**, 6819 (1986).

[181] S. Khare and T. L. Einstein, *Brownian Motion and Shape Fluctuations of Single-Layer Adatom and Vacancy Clusters on Surfaces: Theory and Simulations*, Phys. Rev. B **54**, 11752 (1996).

[182] J. C. Hamilton, M. Sørensen, and A. F. Voter, *Compact Surface-Cluster Diffusion by Concerted Rotation and Translation*, Phys. Rev. B **61**, R5125 (2000).

[183] O. U. Uche, D. Perez, A. F. Voter, and J. C. Hamilton, *Rapid Diffusion of Magic-Size Islands by Combined Glide and Vacancy Mechanism*, Phys. Rev. Lett. **103**, 046101 (2009).

[184] H. Wu, A. Signor, and D. Trinkle, *Island Shape Controls Magic-Size Effect for Heteroepitaxial Diffusion*, Arxiv preprint arXiv:0908.3006 (2009).

[185] A. Signor, H. Wu, and D. Trinkle, *Misfit-Dislocation-Mediated Heteroepitaxial Island Diffusion*, Arxiv preprint arXiv:0908.3004 (2009).

[186] R. Pushpa, J. Rodríguez-Laguna, and S. N. Santalla, *Reconstruction of the Second Layer of Ag on Pt (111): Extended Frenkel-Kontorova Model*, Phys. Rev. B **79**, 085409 (2009).

[187] Y. Nourani and B. Andresen, *A Comparison of Simulated Annealing Cooling Strategies*, J. Phys. A **31**, 8373 (1998).

[188] S. J. Plimpton, *Fast Parallel Algorithms for Short-Range Molecular Dynamics*, J. Comp. Phys. **117**, 1 (1995).

[189] R. Stumpf and M. Scheffler, *Theory of Self-Diffusion at and Growth of Al(111)*, Phys. Rev. Lett. **72**, 254 (1994).

[190] R. Stumpf and M. Scheffler, *Erratum: Theory of Self-Diffusion at and Growth of Al(111)*, Phys. Rev. Lett. **73**, 508 (1994).

[191] P. Sprunger, E. Lægsgaard, and F. Besenbacher, *Growth of Ag on Cu (100) Studied by STM: From Surface Alloying to Ag Superstructures*, Phys. Rev. B **54**, 8163 (1996).

[192] Y. Tiwary and K. A. Fichthorn, *Connector Model for Describing Many-Body Interactions at Surfaces*, Phys. Rev. B **78**, 205418 (2008).

[193] F. D. Murnaghan, *The Compressibility of Media under Extreme Pressures*, Proc. Nat. Acad. Sci. **30**, 244 (1944).

[194] D. Lide and H. Frederiksen, eds., *CRC Handbook of Chemistry and Physics: A Ready-Reference Book of Chemical and Physical Data* (CRC Press, 1995), 76th ed.

[195] J. Rogal, *Stability, Composition and Function of Palladium Surfacs in Oxidizing Environments: A First-Principles Statistical Mechanics Approach*, Ph.D. thesis, Freie Universität Berlin (2006).

[196] L. Liu and W. Bassett, *Compression of Ag and Phase Transformation of NaCl*, J. Appl. Phys. **44**, 1475 (1973).

[197] J. E. Peralta, J. Uddin, and G. E. Scuseria, *Scalar Relativistic All-Electron Density Functional Calculations on Periodic Systems*, J. Chem. Phys. **122**, 084108 (2005).

[198] J. D. Silva, C. Stampfl, and M. Scheffler, *Converged Properties of Clean Metal Surfaces by All-Electron First-Principles Calculations*, Surf. Sci. **600**, 703 (2006).

[199] H. Orita, N. Itoh, and Y. Inada, *All-Electron Scalar Relativistic Calculations on Adsorption of CO on Pt(111) with Full-Geometry Optimization: A Correct Estimation for CO Site-Preference*, Chem. Phys. Lett. **384**, 271 (2004).

Die VDM Verlagsservicegesellschaft sucht für wissenschaftliche Verlage abgeschlossene und herausragende

Dissertationen, Habilitationen, Diplomarbeiten, Master Theses, Magisterarbeiten usw.

für die kostenlose Publikation als Fachbuch.

Sie verfügen über eine Arbeit, die hohen inhaltlichen und formalen Ansprüchen genügt, und haben Interesse an einer honorarvergüteten Publikation?

Dann senden Sie bitte erste Informationen über sich und Ihre Arbeit per Email an *info@vdm-vsg.de*.

Sie erhalten kurzfristig unser Feedback!

VDM Verlagsservicegesellschaft mbH
Dudweiler Landstr. 99   Telefon +49 681 3720 174
D - 66123 Saarbrücken   Fax   +49 681 3720 1749
www.vdm-vsg.de

Die VDM Verlagsservicegesellschaft mbH vertritt

Printed by Books on Demand GmbH, Norderstedt / Germany